"C. S. Lewis once told an audience of Oxford students that God isn't glorified when we tell lies to protect his reputation. Honoring God with our minds means a resolute commitment to the facts. And the facts, as Janet Kellogg Ray explains, tell us evolution is true. That's not only okay for Christians to believe, it's imperative. And it's beautiful. Writing with equal parts humor and clarity, Ray winsomely confronts strawman arguments from creation apologists with accessible explanations about how evolution actually works and what that means for humans created in God's image. A rewarding book for Christians of all ages and education levels who want to understand God's creation in all its complexity."

—Samuel L. Perry
Sam K. Viersen Presidential Professor of Sociology,
University of Oklahoma

"*Fish with Feet* takes us on a lively and engaging tour of the living world and its grand, unifying theory—the theory of evolution. Beautifully written, specifically addressed to people of faith, on every page it confirms Darwin's vision that there is indeed grandeur in this view of life."

—Kenneth R. Miller
author of *Finding Darwin's God*

"*Fish with Feet* walks readers through the basics of evolution without caricature or confrontation. As always, Janet Kellogg Ray writes with clarity, charity, and occasional hilarity."

—Glenn Branch
deputy director, National Center for Science Education

"As Janet Kellogg Ray examines the wide-ranging scientific findings pertaining to human evolution, she exposes the myths and misunderstandings that lead many Christians to believe that evolutionary science poses a challenge to their faith. She does this with humor and empathy, as one who once struggled with these questions and challenges herself, and writes at a level that is broadly accessible."

—Luke J. Janssen
professor emeritus of medicine, McMaster University

FISH WITH FEET

Human Evolution and the Image of God

JANET KELLOGG RAY

WILLIAM B. EERDMANS PUBLISHING COMPANY
GRAND RAPIDS, MICHIGAN

Wm. B. Eerdmans Publishing Co.
2006 44th Street SE, Grand Rapids, MI 49508
www.eerdmans.com

Published 2026
Printed in the United States of America

32 31 30 29 28 27 26   1 2 3 4 5 6 7

ISBN 978-0-8028-8561-6

Library of Congress Cataloging-in-Publication Data

Names: Ray, Janet Kellogg, 1960– author
Title: Fish with feet : human evolution and the image of God / Janet Kellogg Ray.
Description: Grand Rapids, Michigan : William B. Eerdmans Publishing Company, [2026] | Includes index. | Summary: "A science educator explains the compelling evidence for human evolution and explores its implications for people of faith"—Provided by publisher.
Identifiers: LCCN 2025029894 | ISBN 9780802885616 paperback | ISBN 9781467470209 epub
Subjects: LCSH: Human evolution—Religious aspects | Religion and science
Classification: LCC BL263 .R39 2026
LC record available at https://lccn.loc.gov/2025029894

For Tabitha and Austin,

"See, I have written your name on the palms of my hands."

CONTENTS

1

WHAT'S THE PROBLEM WITH MONKEYS?

Bubbles. Orbits. Echo chambers. Cozy little spots where we find ourselves swaddled in the warmth of group-think.

It's easy to think everyone agrees and disagrees along the same lines. It's easy to think that our bubble's opinion is the majority opinion, and of course, the correct one.

It's also easy to shoot down a position with which you disagree if you never consider the position of the actual position-holders themselves. We build a big-ole straw man out of what we *think* the opposition believes, tear it down, and congratulate ourselves on a job well done.

So, on a clear December night a week before Christmas 2022, I sat in a crowd of about two hundred to hear—firsthand, from a premier source—why rejection of human evolution is both scientific *and* the only reasonable position for people of faith.

The event was the debut of *Adam or Apes*, a new film from the Institute for Creation Research (ICR). Dallas-based ICR has beautiful facilities and produces loads of professional media. An on-site Discovery Center is a museum-style venue—a popular destination for homeschool and Christian school field trips. The film screening was followed by a panel discussion and a question-and-answer session by ICR staff.

I was there to listen and learn, so I didn't ask a question during the Q&A time. My goal was to hear the best arguments against human evolution from the evolution-rejecting precinct of Christianity. Instead, I heard two hours of counterarguments about evolution that *no one* is making. In 2022, ICR's argument was still about Piltdown man, a forgery exposed more than seventy years ago—and by the way, exposed by scientists.

Curious to find out if ICR's arguments against human evolution had evolved, I returned to ICR's Discovery Center two years later to tour the exhibits. The center features Christian history, Noah's ark, the Tower of Babel, and exhibits about the age of the earth, but my focus on this visit was human evolution.

I tried not to be drawn offsides by the 360-degree opening video with its blond, fair-skinned, blue-eyed Adam sporting a 2024 haircut and trendy facial scruff.

First up were poster-style exhibits about human DNA. Humans and chimpanzees do not share 98 to 99 percent of their DNA, as "secular" scientists claim; it's only 84 percent.[1] In the second poster, we learn there is no such thing as pseudogenes.[2] Both posters are unambiguous: humans share no common ancestry with primates.

Next up—the fossil record. In an exhibit titled "Man or Myth?," three examples are presented: (1) the famous Lucy fossil, (2) extinct people groups like the Neanderthals, and (3) ICR's favorite, Piltdown man. Another exhibit features seven famous fossils who, according to "secular" scientists, are human ancestors. More than twenty hominin species have been identified to date, yet one example of the seven in this exhibit is—you guessed it—Piltdown man.

It's easy to cast doubt on human evolution if you think "secular" scientists are still fooled by Piltdown man. Instead of presenting peer-reviewed evidence dismantling evolution or peer-reviewed evidence for special creation of humans, ICR built a straw man out of bogus evolution arguments and dismantled it.

DARWIN THE MONKEY-MAN

Human evolution is, no doubt, a touchy topic. Charles Darwin's masterwork, *On the Origin of Species*, barely hinted at the topic. Darwin postponed publishing *Origin* for years, telling a trusted friend that proposing evolution—in general—felt like "confessing to a murder." Darwin closed *Origin* with a vague reference to future study: "Light will be thrown on the origin of man and his history."[3]

Twelve years later, Darwin followed *Origin* with *The Descent of Man*. Darwin never believed or wrote about monkey-into-human evolution. His human argument, as it was with his broader theory, was shared common ancestry.[4] Still, popular culture glommed onto the monkey-man theme and took off. French caricaturist Faustin Betbeder's most famous sketch depicts Darwin as a monkey with a human head, holding a mirror before an actual monkey seated next to him. The sketch is captioned with a quote from Shakespeare's *All's Well That Ends Well*: "Some four or five descents since."

A few years later, Darwin was featured on the cover of the French satirical magazine *La Petite Lune*. Again,

Darwin is a monkey, this time hanging from a tree branch with one hand and twirling a long monkey-tail in the other.

The French didn't think much of Darwin, to say the least. First (and probably foremost), Darwin was English, and that was an automatic strike against him from the start. On top of that, thoroughly Catholic France saw Darwin and his theory as displacing God and upsetting traditional morals.

So of course, when two bumbling criminals committed a grisly murder in Paris in 1878, Charles Darwin was blamed. Oh, Darwin had no part in the crime, but investigators discovered that one of the murderers had delivered a public lecture praising Darwin's theory just three weeks after the murder. The conservative French establishment ran with it: If humans are no more than apes, heinous crimes can be excused.[5]

SHOWDOWN IN MONKEY TOWN

The 1925 Scopes monkey trial was considered the "Trial of the Century," at least until the O. J. trial dethroned it. The Scopes trial was a setup from the beginning, an alliance between the business leadership of Dayton, Tennessee, and the American Civil Liberties Union (ACLU). Town leadership wanted the economic boon of a high-profile trial, and the ACLU wanted a verdict that could be appealed to the Supreme Court of the United States.

The Scopes trial wielded the monkey theme for both comic and evidentiary effects. It was July and sweltering inside the courtroom, so the action was moved outdoors. The atmosphere was more fairgrounds than hallowed grounds, with vendors selling monkey souvenirs, little girls playing with monkey dolls, and a suit-and-fedora-wearing chimpanzee named Joe Mendi casually sipping a Coca-Cola.

The monkey theme was a bit of a curiosity for a trial centering Darwin. Charles Darwin was far more obsessed with the evolution of beetles, barnacles, and pigeons than he was with the evolution of primates and humans. But no one sold barnacle dolls, and no one dressed up a pigeon and bought it a Coke.

In the Scopes trial, monkeys weren't simply for comic effect. John Scopes, the high school teacher/defendant in the trial, was *not* accused of teaching evolution. Scopes was accused of violating Tennessee's Butler Act, which made it unlawful to teach any theory that "denies the story of the Divine Creation of man as taught in the Bible, and to teach instead that man has descended from a lower order of animals."[6]

Monkeys put John Scopes in the hot seat. It wasn't beetles or barnacles or earthworms or even giant extinct glyptodonts, all of which fascinated Darwin. Apparently, the state of Tennessee was okay with the evolution of those things, but monkeys and the implication of human evolution were a bridge too far.

A MODERN MONKEY TRIAL

Eighty years after Scopes, a new evolution trial in a new century again put monkeys in the spotlight. The local school board in Dover, Pennsylvania, planned to adopt an intelligent design creationist textbook as an "option" to evolution in high school biology. Federal Judge

John E. Jones III, a conservative Republican appointed to the bench by George W. Bush, ruled in *Kitzmiller v. Dover* that intelligent design is a religious construct and, as such, has no place in a public-school classroom . . . but not before the local community spiraled into turmoil.

When a group of parents (as well as the entire science faculty at the high school) opposed the board's plan, things got ugly. Prior to the trial, there were lots of overwrought public discussions, as you can imagine. At one such event, school board member William Buckingham shouted down critics of the proposal by evoking Jesus and his death on the cross: "Can we have the courage to stand up for him?" pleaded Buckingham.[7] Evolution, according to Buckingham, is not only wrong—it also makes Jesus sad.

The similarities of the Dover trial to the Scopes monkey trial were obvious. Like the original trial, the broader topic of evolution narrowed to monkeys and revulsion toward human ancestry with such beasts. Prior to the trial, local pastor Jim Grove noted, "It's a monkey trial all right. And the evolutionists are the monkeys."[8] And during a break in trial testimony, board member David Napierski quoted his young daughter: "One thing's for sure, we didn't come from any monkey. She knows that as well as I do."[9]

Outside the courtroom, things got really ugly. Although Dover high school students could leave the classroom when the assistant superintendent read the required statement about "options" to evolution, few students left. The entire science faculty stood in the hall in solidarity with the opt-outs. The fourteen-year-old daughter of Tammy Kitzmiller (whose name was given to the lawsuit) endured taunts of "Hey, Monkey Girl!" at school.[10]

As with the Scopes trial, monkeys ruled the day among those opposing evolution. Fortunately, the expert witnesses at the Dover trial maintained focus on the question at hand: Can creationism be taught as an alternative to evolution in a public school? And thankfully, science won the day.

THE MONKEY PROBLEM

For people with a reluctant acceptance of evolution in general, human evolution is often where they get off the bus. Opposition to human evolution is rarely about evidence. More often, it's a visceral reaction to a perceived insult.

Newspaper reporter Lauri Lebo sat through every day of testimony in the *Kitzmiller v. Dover* trial. Her father was appalled at Lebo's acceptance of evolution, as were local pastors—but it wasn't the evolution of flowering plants or even four-limbed land animals that upset them. It was the humans: "Teaching the traditional Darwinian concept that man evolved from lower forms of life, that's almost a slap in my face that takes away the dignity from humanity," said Pastor Ray Mummert.[11]

Opposition to human evolution is rarely about evidence. More often, it's a visceral reaction to a perceived insult.

Unlike the evolution of any other organism or the workings of any mechanism of the evolutionary process, human evolution is distinctly offensive.

Fazale Rana, president and CEO of Reasons to Believe, accepts an ancient age for the earth but not the ancient evolution of humans: "Instead of regarding humans as 'fearfully and wonderfully made' (as Scripture teaches), the evolutionary paradigm denigrates human beings *as a logical entailment of its mechanisms*. It renders human beings as nothing more than creatures."[12]

A twelve-year study of university students in an introductory anthropology course likewise found creationist students repulsed by the thought of humans as animals: "God created humans to be Christ-like and animals to be just that. Nothing more. Nothing less. No relation to me!!" And of course, the most offensive animal to be is a monkey: "God created Adam and Eve not a monkey or some other animal."[13]

Eric Hovind founded the Creation Network, an organization boasting worldwide reach through websites, films, and hundreds of episodes of the *Creation Today* talk show. Hovind gets right to the point: Teaching evolution convinces children that life is meaningless. "If you tell kids they came from animals, don't be surprised when they act like animals," says Hovind.[14]

HUMAN EVOLUTION—NO THANKS

Both Pew and Gallup polls have tracked American belief in human evolution for more than a decade. Until recently, close to half of Americans believed that God specially created humans "in their present form," with no evolution involved. Currently, the number of Americans who reject human evolution is around 35 percent.[15]

In the only nation to put humans on the moon, why do people reject the science of how those humans evolved to do so? What are we missing?

We need to go back to school—specifically, high school biology class. Even in public schools, human evolution is a hot-potato topic. Wanting to (understandably) avoid controversy and the ire of parents and community, 60 percent of public-school biology teachers soft-pedal the evolution chapters.[16] Soft-pedaling usually looks like either (1) downplaying the importance of evolution ("just learn it for the test") or (2) focusing on change at the molecular level and avoiding species-level change.

What many biology teachers avoid at all costs is the human part. Glenn Branch, deputy director of the National Center for Science Education, pulls no punches; creationists, he says, "are not invested in whether evolution affects the sizes and shapes of the beaks of finches in the Galápagos, they are worried about whether people were created in the image of God himself."[17]

Currently, almost half of public high school biology teachers in Arkansas allocate little to no instruction to human evolution. These same teachers willingly taught other aspects of evolution—like natural selection and speciation—but ignored the humans.[18]

Even in a science-friendly audience, human evolution can be a touchy topic. *Recovering Evangelicals* is a popular podcast focusing on the tension between Christianity and the twenty-first century. Host Luke Janssen is a physiologist, and his topics, his cohost, and his audience are overwhelmingly friendly to the science of evolution. For a year or so, the podcast explored various topics in evolution—and the audience grew steadily. But audience numbers took a steep dive during the two weeks Janssen discussed human evolution, and it took months for the numbers to recover.[19]

Monkeys are a problem. Common ancestry with apes is offensive and demeaning, and we just aren't having it.

CHAPTER 1

FROM THE BEGINNING

Human evolution is far knottier than a simple "did we come from monkeys or not?" The human story penetrates the deepest nooks and crannies in the history of life.

We'll start at the beginning. It's an incredible journey to us. And the story is far more captivating than the mockery of a chimp birthing a human baby. The story involves lots of moving parts and pieces and eons of time.

"If humans came from monkeys, why do we still have monkeys?" is often the ace of spades thrown down on the evolution card table. Without a doubt, humans are complex, and humanity is complicated. It's hard to imagine progressing from goo in a warm little pond to Marie Curie without a micromanaging designer/creator. The story of humans is deep and rich and precarious—and immensely captivating.

My adult kids have precious nicknames for me, like "the Questionnaire" and the "Questioning Beast," due to my propensity for rapid-fire questions when I want information. I come by it naturally—I've gone into the wee hours of the morning discussing evolution with my eighty-eight-year-old dad. No matter how complete (so I thought) my answer, he counters with "well, what was before that?" This book is an attempt to answer the question "what was before that?"

The answer to "what was before that" regarding human evolution is unfortunately informed by a very *ill*-informed graphic. Based on a 1941 illustration popularly known as "March of Progress," the image features a left-to-right, all in a row, one behind the other progression from a tailed monkey to a tail-less ape, followed by a couple of ape-ish/man-ish creatures, then a caveman guy, and finally, the line leader—a modern upright human. In reality, we did not progress *from* them, says evolutionary biologist Prosanta Chakrabarty; "rather, we progressed *with* them from our shared ancestor."[20]

THE NAME GAME

In the eighteenth century, it was a challenge to puzzle out which animals or plants were related to each other. Natural scientists named things as they willed, according to their own observations and creativity, and in their own language. As you can imagine, a bird or insect could have a multitude of names given by observers in disparate locations.

In 1735, Swedish botanist Carl Linnaeus devised a uniform system for naming living things. Linnaeus organized animals and plants into nested hierarchies and gave each level, from most inclusive to the individual, a consistent Latin name. If you memorized a mnemonic like "King Phillip Came Over From Great Spain" to recall classifications in biology class (kingdom, phylum, class, order, family, genus, species), you can thank the good Swede.

Using Linnean taxonomy, a rat is a rat, but it is also a rodent, a mammal, a vertebrate, and an animal, all at once. The system was a nice and neat nested arrangement in which a rodent didn't stop being a mammal when rodents branched off the mammal family tree.[21]

Linnaeus organized things according to apparent similarities. You have wings and you fly? You're a bird.

You're a mammal but you look like a fish and live in the ocean? You're a whale. Interestingly, Linnaeus grouped humans with monkeys and apes, but not because of evolution—it would be another century before evolution entered the conversation. Linnaeus placed humans with monkeys and apes because of their similar anatomy. Before Darwin, before Scopes, and long before modern genetics, Linnaeus recognized this fact: We are apes—*Homo sapiens*.

Modern classification (called systematics) still utilizes hierarchies but organizes living things according to their evolutionary history: Who is related to whom, and how are they related? A *clade* consists of a common ancestor and all its descendants. Clades can form within existing clades—we find clades within clades within clades—a nested hierarchy. Once you are in a clade, you are always part of that clade.

You can't evolve out of a clade. You can go extinct, but you are an extinct member of that clade. Once you're in, you're in. It's kind of like the Hotel California—you can check out, but you can never leave. Descendants of vertebrates will always be vertebrates. Descendants of reptiles will always be reptiles—that's why birds are biologically reptiles. Descendants of fish will always be fish. Apes are one of the many clades that branched off the lobe-finned fish clade.

You are an ape. You are also a fish. We are a lot of other things, too, but for the purposes of this book, fish and apes are defining points in our evolutionary history.

Evolution is not a march of progress. It is a tree, but it's not a tree with well-defined and clearly marked branches. It is a bushy tree with smudgy demarcations.

Scattered throughout this living tree are some very ancient branches, so old it's amazing they are still going strong. Scattered throughout the tree are dead branches, once living, no longer living, no longer growing, but still connected and found in place among the branches to which they are most closely related.

We live in an amazing world of insects, plants, fungi, bacteria, and all things weird and wonderful. But for now, our focus is us. This book is self-centered.

A WORD ABOUT EVOLUTION

This book is not a defense of evolution. We begin the journey with this reality: Evolution theory is the foundational principle of all modern biology. Imagine trying to understand chemistry without atoms (atomic theory) or medicine without germ theory. Evolution theory forms the scaffolding for our research and understanding of modern medicine, anatomy, physiology, agriculture, genetics, conservation, and ecology.

Evolution theory explains why the facts we observe in biology are the way they are. Evolution theory explains why the laws we observe in biology hold true. You may not accept evolution, or you may accept evolution, or you may be somewhere in between. Acceptance or rejection doesn't change the fact that evolution forms the scaffolding for understanding modern biology. Acceptance or rejection doesn't change the fact that evolution theory is, and continues to be, a powerful predictor of new knowledge.

Absolutely there are people who accept evolution and acknowledge God as the originator of all that is. They acknowledge God as sustainer and the underlying wisdom of the universe. I am one of them. Some people with these beliefs self-identify as "evolutionary creationists" or "theistic evolutionists." I don't—I simply describe myself as a person of faith who accepts the science of evolution. If I was a weather forecaster, I wouldn't describe myself as a "theistic meteorologist."

In this book, I use the terms "creationist" and "creationism" to describe a specific viewpoint about the origins of life and the universe. In general, creationists do not believe that natural processes alone can explain the origins, development, and diversity of life. In general, creationists believe that, to some degree, God micromanaged the process. Creationists are "young earth" creationists (the universe is six thousand to ten thousand years old) or "old earth" creationists (the universe is ancient). Some creationists prefer the term "intelligent design," but design is not a belief separate from creationism. Rather, intelligent design is a way of explaining creationism.

Whether you accept evolution or not doesn't change the fact that it is the foundational principle of all fields of modern biology. Given that fact, let's go on a journey of human evolution. The story of us.

DISCUSSION PROMPTS

1. Which of the following statements best describes your view of human origins? Is this a long-held view or a more recent conclusion? Would a modified version of one of these fit you better? If so, how would you modify the statement?

- Humans have existed in their present form since the beginning of time.
- Humans evolved, by natural processes, over time.
- Humans evolved by natural processes, guided by God.

2. Regardless of your answer to the first question, respond to these two questions: In your opinion, what is the best evidence or argument *for* human evolution? In your opinion, what is the best evidence or argument *against* human evolution?
3. On a scale of one to ten, how would you rate your degree of knowledge and understanding of the evidence for human evolution? What kinds of resources informed your understanding? Church? School? The Internet? Friends? Books or other materials? A combination of resources?
4. July 2025 was the 100th anniversary of the Scopes monkey trial. John Scopes did not break the law by teaching evolution; he broke the law by teaching *human* evolution. Did you learn about human evolution in school? Are you aware of any efforts in your public-school district or by your state school board to curtail the teaching of evolution? Are you aware of any efforts to teach "strengths and weaknesses" of science topics deemed "controversial" by some people? Do you have an opinion regarding such efforts?
5. Many people who accept evolution for bacteria, viruses, plants, and nonhuman animals draw the line at human evolution. Why might that be so?
6. Eric Hovind said, "If you tell kids they came from

animals, don't be surprised when they act like animals." What are your thoughts about Hovind's statement? Is it a reasonable assessment? Why or why not?

7. "If we came from monkeys, why are there still monkeys?" is an argument often heard in evolution discussions. What is the point of this statement? What are the implications of this statement?
8. Linnaeus was the first to systematically organize living things into nested hierarchies. Using Linnean taxonomy, explain why a rat is a rat, but a rat is also a rodent, a mammal, a vertebrate, and an animal—all at once.
9. A "clade" is a common ancestor and all its descendants. Modern classification organizes life using clades. An organism cannot evolve out of a clade, even if it is extinct. Explain.
10. Do you consider yourself a creationist? Why or why not? If you're not sure, how might you describe yourself? Would you modify the term? Would you use another term?

2

FROM GOO TO YOU BY WAY OF THE ZOO

I try to behave myself in church, but I'm unfortunately afflicted with audible eyerolls. The sermon was about the wonder of the human body. We are "fearfully and wonderfully made" was the theme. We're complicated and unexplainable.

The preacher opened with a quote from Bill Bryson's *A Short History of Nearly Everything*: "by all the laws of probability, proteins shouldn't exist."[1] I just *knew* what was coming next—an analogy to illustrate protein construction, told with incomprehensibly large numbers and followed by the slam-dunk of minuscule chances of it ever happening.

Proteins are built of amino acids, strung together like beads on a string. Once strung together, the string bends, corkscrews, and spirals into a three-dimensional shape, giving a protein its unique function.

Collagen is an abundant and vital protein in your body, and by odds (according to the sermon analogy), shouldn't exist. To build a molecule of collagen, 1,055 amino acids must line up on the string, and in an exact sequence. Without micromanagement (the analogy implies), the chances of getting all 1,055 amino acids in the correct position are astronomical.

Bryson takes us to Las Vegas to explain the impossibility of it all. Imagine a slot machine, ninety feet long, with 1,055 spinning wheels, each representing one amino acid in the collagen chain. How many times must you pull the lever before all 1,055 come up at once, and in the correct order? Incomprehensible. Let's make it easier—substitute a shorter protein, say 200 amino acids long. The number of pulls on the lever to get all 200 in the correct order would be larger than all the atoms in the universe.

The "747 in a tornado" is a popular remix of the same analogy. In this analogy, all the bits and pieces and odds and ends of a 747 jet are present in a junkyard. The sky darkens, and an F5 tornado drops down and moves through the junkyard. The result? A fully assembled, ready-to-fly airplane.[2]

And here's a fun one: a person on the International Space Station shoots a gun at a target on Earth—a target with a bull's-eye only one trillionth of a trillionth of an inch in diameter—and hits the bull's-eye perfectly.[3] Anytime there is a suggestion of life arising from nonlife or self-assembly of the complex chemicals needed for life, big numbers and impossible odds are employed to challenge the claim.

That settles it, right? Life cannot come from nonlife. Life cannot arise spontaneously. The odds are stacked against it and can't be overcome.

Not so fast. The "proteins are impossible" argument is a common one, but Bryson's narrative takes a surprise turn.

Unfortunately, the preacher stopped with the "impossible" quote. Just a few paragraphs later, Bryson takes us back to Vegas and that comically long slot machine. What if 199 wheels or even 1,054 wheels were already locked in place, and we only needed the last symbol in the sequence? Suddenly, the odds change. What if, Bryson suggests, proteins did not assemble all at once? What if complex proteins were assembled from smaller, less complex components?

What if, says Bryson, proteins *evolved*?

Now let's do literary monkeys. One million monkeys, each with his own typewriter, pecking away randomly, with one assignment: produce a single sentence. Specifically, type out "to be or not to be" from *Hamlet*—no capital letters or spaces required. Settle in . . . it would take seventy-eight thousand years for our little monkey typists to produce "tobeornottobe."

Richard Hardison of Glendale College wrote a computer program in the 1980s, mimicking the famous million monkeys and their typing skills. Every time a monkey happens to type the correct letter in the correct position, the letter remains. In under ninety seconds, the computer program produced the correct phrase. In only four and half days, the program produced *Hamlet* in its entirety.[4]

Smart monkeys.

THE CHEMISTRY OF LIFE

All life is chemistry. Everything that happens in your body—all of it—is, at its foundation, a chemical reaction. Nerves fire, muscles contract, blood sugar is regulated, the brain interprets images from the eyes and sounds from the ear . . . all of this happens at the molecular level.

Four complex macromolecules—proteins, lipids, carbohydrates, nucleic acids—build, regulate, protect, defend, and reproduce our bodies.

Molecules form when atoms bond together, but here's the bottom line: It's not random. There's a deep structure to life—physics and chemistry place constraints on biology.[5] Biological problems can be solved in only a few ways. The forces that bind at the molecular level vary in strength and circumstances and are dependent on the physical makeup of the atoms themselves. Molecule formation can be complicated, but it's not random.

Simple molecules unite to form more complicated molecules. Complicated molecules bend and twist and fold, creating new bonding options and a myriad of functions made possible by their three-dimensional shape.

When a monkey types two correct letters, we lock those in, and the odds of getting the correct phrase improve. Once molecular bonds are formed, there is a finite number of other bonds that can be made.

NO SOUP FOR YOU

Life from nonlife (abiogenesis) is collectively scorned by those who reject a natural explanation for the beginning of life. Molecules to man? Ridiculous. Rocks to rabbits? Get real. Life is too complicated and too complex to arise accidentally in a pool of muck. From goo to you by way of the zoo is simply an atheist's fairy tale.

In public, Charles Darwin generally stuck to describing the evolution of life, once life began. He tiptoed around the topic of *how* life began, barely touching on it at the end of *On the Origin of Species*. But in a private letter to a friend, Darwin was far chattier: "But if (and Oh! what a big if!) we could conceive in some warm little pond, with all sorts of ammonia and phosphoric salts, light, heat, electricity, etc., present, that a protein compound was chemically formed ready to undergo still more complex changes."[6]

Ah-h-h . . . Darwin's "warm little pond." In the safety of a private letter, Darwin speculated about the beginning of life in a primordial chemical soup.

Abiogenesis, or the chemical evolution of life, was put to the test in 1953. In an iconic experiment, University of Chicago researchers Stanley Miller and Harold Urey filled a glass tube with gases thought to represent the atmosphere of the early earth. Miller and Urey added water and shot the whole thing through with sparks of electricity.

The result was a dark broth, filled with organic compounds and containing five different amino acids—the building blocks of proteins. Fifty years later, Miller's former student, Jeffrey Bada, reanalyzed some of Miller and Urey's leftover soup. Using more sensitive equipment than was available in the 1950s, Bada found all twenty amino acids present in modern organisms.[7]

Miller and Urey's original experiment has been modified in various ways over the decades. We know more about the earth's early atmosphere than we did in the 1950s, so we've experimented with different gases, energy sources, and environmental conditions.[8]

We don't, however, have the exact recipe for primordial soup. Should we throw up our hands with a soup-Nazi growl and bark "no soup for you!"? Not at all. Although we don't know the exact conditions present on a billions-of-years-old earth, we have a pretty good idea.[9]

COOKING THE SOUP

The potential for soup is one thing, constructing the soup is another. Miller and Urey used electricity (simulating lightning strikes) to energize their reactions. Instead of hit-or-miss input of energy from lightning, what if life arose in an environment with a constant source of energy?

Deep below the surface of the ocean, a jet of black smoke and superheated mineral-rich water shoots out of a vent in the sea floor, in eighteen-story-tall plumes. Like the geysers of Yellowstone and those erupting across Iceland, hydrothermal vents on the ocean floor spew water at temperatures greater than 100 degrees Celsius—higher than the boiling point. Under extreme pressure at such depths, the water stays in a liquid state.

Deep-sea hydrothermal vents were first discovered in 1977 during exploration of the Galápagos Rift in the eastern Pacific. And what a discovery: never-before-seen ecosystems and hundreds of new-to-us species. Despite extreme temperatures, toxic minerals, and a complete lack of sunlight, communities surrounding the vents were thriving.[10]

Apparently, these exotic creatures skipped science class on photosynthesis day. Something else is powering the communities.

Deep-earth water rushing out of the vents can be very acidic, with a pH much lower than the surrounding seawater. And voilà! we now have a concentration gradient—an ordinary, everyday process used by plants, animals, and bacteria to fuel life's activities.[11]

A concentration gradient is formed when there is a higher concentration of something on one side and a lower concentration on the other side. Things move toward the area of lower concentration until everything is equal. In the case of deep-sea vents, the disparate pH levels between the vent water and the surrounding seawater create a concentration gradient.

Concentration gradients do not require energy to work. In fact, they *generate* energy. Think of a hydroelectric dam with a large concentration of water behind a retaining wall. Open a small port in the wall, and water rushes toward the side with no water with so much force that turbines spin and electricity is generated.

Concentration gradients in your cells power the conversion of food into energy for your body. Concentration gradients power the synthesis of sugars in plants. In a simple, elegant process, life harnesses nonlife to fuel life.

We do not know the exact recipe for the primordial soup that formed and fueled the nurseries of life, but this we know: It's possible.

The water spewing from the mouth of an ocean-floor vent also carries a boatload of minerals, picked up from deep within the earth. As the water exits, mineral-rich debris collects around the mouth of the vent.

We do not know the exact recipe for the primordial soup that formed and fueled the nurseries of life, but this we know: It's possible. The experiments of Miller and Urey and experiments in the decades that followed demonstrated that amino acids—the very building blocks of proteins—can be synthesized from inorganic substances.

In the dark depths of the oceans, mineral-rich deposits aggregate in the presence of a constant source of electric power created by a pH gradient. The rocks at the mouth of the vents are pocked with microscopic pores, just the surface needed to catalyze organic reactions.[12]

Life didn't just "pop" into existence but likely emerged from a series of chemical reactions. Simple molecules are always forming more complicated molecules. It's chemistry. It's electrical forces. It's predictable.

Biological anthropologist Eugenie Scott noted the difficulty of drawing an unambiguous line between life and nonlife: "It is beginning to look like the origin of life was not a sudden event, but a continuum of events producing structures that, early in the sequence, we would agree are not alive, and at the end of the sequence, we would agree were alive, with a lot of iffy stuff in the middle."[13] Russ Miller, a young-earth-creationist author and popular speaker, mocked Scott for admitting gaps in our understanding of abiogenesis: "a lot of iffy stuff in the middle? That's the modern 'college' explanation of how life got started without God."[14]

The most primitive life-forms on Earth have a chemical makeup that suggests they originated in a superheated environment, well above boiling.[15] Deep-sea vents have all the ingredients and the required energy and are at present the most likely locations for primordial nurseries.

We don't know everything about abiogenesis, but that doesn't mean we know nothing.

ARRIVAL OF LIFE

Organic molecules, tucked inside some kind of slime or a fatty bubble, could go about their very simple chemical ways, protected just a bit from the turbulent outside world. Orderly construction was possible: component parts reacting and binding together to form larger molecules. With a wall—ever so thin—separating inside from outside, little personalized concentration gradients could generate energy to fuel it all.

Inside whisper-thin little havens of protection, the "iffy stuff" was busy. What was the project at hand—building self-replicating molecules, the precursors of DNA and RNA? Or was it simple yet coordinated chemical reactions, the forerunners of complex metabolisms? Or was it both?

For the first half-billion years or so after its formation, the earth's surface was unstable and churning with molten rock—this period in Earth's history is not called the "Hadean" for nothing. But as soon as the earth cooled, we find evidence of life. Life, at its chemical basis, is made of the most common stuff: carbon, oxygen, hydrogen, nitrogen, sulfur, phosphorus. Life is not made of the rare or weird.

According to The Bump, a website for expectant parents, single-syllable baby names are trending. Some years the lists are populated with modern concoctions like McKinleigh; the next year there's a vintage takeover and Vera tops the charts. The most popular name 4.2 billion years ago was a gender-neutral "LUCA"—gender-neutral not because it was trendy but because there was not yet such a thing as gender. What's more, it was the only name on the list.

Meet LUCA, the Last Universal Common Ancestor of all modern life. And no, he was not traded to the Los Angeles Lakers.

LUCA probably was *not* the original living cell. Most likely, life arose in fits and starts, with many dead ends. LUCA was simply the one that survived. Natural selection has always been the name of the game, and for whatever reason, LUCA won. Everything alive today can trace its ancestry back to LUCA.

What was LUCA like? Surprisingly, we have a fairly good idea. Genes common to all modern life, from bacteria to bison to people, originated with LUCA. Combing through genomes of the most ancient extant lineages, geneticists have identified a large group of genes likely originating in LUCA, coding for about 2,600 proteins. Among these are genes that allow an organism to use hydrogen gas and carbon dioxide for energy and a group of genes known to chop up invading viruses.[16]

LUCA had a gene that protected it from ultraviolet light, suggesting it could have lived on the water's surface. But there's also evidence that LUCA produced an enzyme commonly found in microbes living in extremely high temperatures. Sound familiar? Such is the environment of deep-sea vents, quite possibly the nurseries of life.[17]

THE PROKARYOTES

All cells belong to one of two categories: the prokaryotes or the eukaryotes. Prokaryotic cells are the oldest cells, predating eukaryotic cells by two billion years.

Prokaryotes are all microscopic and are very simple in structure and function. Bacteria are the best known of this group and live in almost every environment on Earth, from the Arctic to oceans to the human gut.

The archaea are also prokaryotes, and likely you've never heard of them—they are the oddballs in the cellular family tree. They live in temperatures above the boiling point, in environments more acidic than vinegar and more basic than ammonia. The archaea have a hard-core nickname: "the extremophiles." Why is this not the name of a metal band?

The oldest life that everyone agrees is "life" is a 3.4-billion-year-old fossilized reef in Australia. Slimy mats of bacteria were buried by sand, followed by another slimy mat and more sand, over and over, forming giant mounds that eventually fossilized. These mounds, called stromatolites, still form today in parts of Australia.[18]

Somewhere between 2.4 and 2.1 billion years ago, a type of bacteria called cyanobacteria started using energy from the sun to make sugar. The sun fueled the process, the sugars fueled the bacteria, and the cyanobacteria flourished. Photosynthesis takes the stage.

Photosynthesis was a groundbreaking innovation, but in addition to sugar, it produced a deadly (at the time) gas as a by-product: oxygen. As the photosynthesizing cyanobacteria spread across the planet, they poured oxygen into the air.

The first cells arose in a world devoid of oxygen. For the earliest organisms, oxygen was a death sentence.[19] The Great Oxidation Event, the first major extinction event in earth's history, killed as much as 99 percent of life on earth.[20]

A few lineages of bacteria survived by adapting to environments without oxygen—these are the anaerobes. Modern anaerobes include the bacteria that cause tetanus and gas gangrene.

In a world bereft of oxygen, metabolisms were slow and low. The survivors of the Great Oxidation Event were those who could use oxygen to drive a ramped-up metabolism. A world filled with oxygen set the stage for complex life-forms with high metabolisms—like us.

THE EUKARYOTES

Back in the day, teachers might have instructed their students to prop up makeshift barriers made of folders or textbooks on their desks to prevent cheating on exams and ensure that students do their own work. A mechanical engineering professor in the Philippines, however, took the practice to a whole new level. Students were instructed to attend the midterm exam with self-designed "anti-cheating" hats.

One student walled off his entire head with egg cartons. Another put tubes over his eyes to give himself literal tunnel vision. There were wigs, helmets, and elaborate character heads. Her students definitely understood the assignment.[21]

Whether it's egg cartons or a propped-up spelling book, the principle is the same: carry out your work without interference from your neighbor.

Around 2.5 billion years ago, a cell intent on "doing its own work" without outside interference made its debut. The eukaryotic cell was a game changer, and life on Earth would never be the same.

For the first two billion years of life, prokaryotic cells were not much more than microscopic bags of fluid with a loop of free-floating DNA sloshing about. Chemical reactions within the cell were constantly adulterated by other nearby reactions. Nothing was concentrated; everything was diluted. In prokaryotic cells, nothing gets done very fast or very efficiently.

Eukaryotic cells are also filled with fluid, but within that fluid are found various structures called organelles. Eukaryotic DNA is linear and is tucked safely inside a membrane-bound organelle called the nucleus. Other organelles are devoted to specific tasks like energy production, protein building, and waste disposal.

Each organelle, working within its own membrane, carries out chemical reactions without interference from nearby reactions. When concentrated in a smaller volume, the rate and efficiency of chemical reactions also increase. And for good measure, potentially harmful molecules are isolated, protecting the rest of the cell.

Membrane-bound organelles allow more complex chemical processes, all done with more safety and efficiency. More efficient and complex chemistry allows for more complex cells, and with complexity comes the ability to communicate and cooperate with other cells.

Eventually, eukaryotic cells would build animals and plants and mushrooms, all made of trillions of eukaryotic cells, all working together.

THE STRANGER WITHIN

How, then, did complex eukaryotic cells evolve from the much simpler prokaryotic cells? How did eukaryotic cells with membrane-bound organelles evolve from prokaryotic cells where everything just sloshes about, willy-nilly? The answer lies in a fundamental strategy for success: teamwork.

Mitochondria are organelles responsible for harvesting energy from food and converting it into a cell-friendly form of energy. The more active the cell, the more mitochondria are found within—muscle cells and brain cells abound with mitochondria; fat cells have few.

Originally, scientists thought mitochondria were bacteria that had somehow infected the cells they were studying. Many in the nineteenth century found this suggestion laughable, but in the 1960s, biologist Lynn Margulis found evidence that this was exactly what happened.[22]

The invasion was not recent, however.

About two billion years ago, a prokaryotic cell (probably one of the archaea) engulfed a smaller bacterial cell.[23] Did the larger cell see the smaller bacterial cell as lunch? Or did the bacterial cell attack and invade the larger cell? We don't know.

But this we do know: For some reason, the engulfed/invading bacterial cell was not digested by the larger cell. We also know that the engulfed/invading bacterial cell had a special skill set—it was particularly good at using oxygen to extract energy from food.

Thus began a mutually beneficial relationship, still going strong to this day. "Life did not take over the earth by combat, but by networking," said Margulis.[24] The engulfed bacteria (now known as mitochondria) benefited from the nutrition provided by the larger cell, and the larger cell benefited from the energy produced by the bacteria. Mitochondria are friendly little hitchhikers who pay for the gas.

The mitochondria found in modern eukaryotic cells are the descendants of free-living bacteria who checked in as permanent houseguests of larger cells, two billion years ago. Here's how we know:

- Each mitochondrion has its very own little chromosome made of DNA, but not a linear chromosome like we see in modern eukaryote cells. Mitochondrial DNA is a circular loop—just like the DNA of modern bacteria. There are only a few genes in mitochondrial DNA, but they are critical for functioning. Mutations in mitochondrial DNA can result in devastating mitochondrial diseases.
- Mitochondria divide independently of the cell in which they reside and do not divide on the same timetable as the cell itself. And when they do divide, mitochondria do not divide in the more complicated fashion of eukaryotic cells. Instead, mitochondria simply pinch in two, just like modern bacteria.
- Mitochondria can only come from other mitochondria. If all mitochondria are removed from a cell, the cell can't build new ones from scratch.
- And like the other organelles of a eukaryotic cell, the mitochondria are enclosed in their own membranes, membranes quite like bacterial membranes.

Other eukaryotic organelles like the nucleus most likely have similar origins.[25]

Lynn Margulis's contribution to our understanding of eukaryotic cell evolution goes beyond the origin of the mitochondria. Margulis was a fiery, spirited soul. She was one of the first to challenge conventional understandings about evolution, and she faced resistance. Thanks to her research, we now understand that evolution can also come about by the sharing and fusion of genomes between species.

Darwin wasn't wrong. Darwin was an early pioneer. Others, like Margulis, refined and reworked the theory. That's how science works.

RISE OF MULTICELLULAR LIFE

I was in Target and in a hurry, and apparently, I couldn't be bothered to put my phone away. I set my phone down *somewhere*, but fortunately I realized it before I drove off. So, there I was, back inside and in line at customer service, spending the time I'd "saved."

Lost Items Guy, opening the lost-and-found drawer: "What does your phone cover look like?" Me (in my head): "Precambrian Ediacaran fauna." Me (out loud): "It's weird." Clearly, mine was the only weird orphaned phone because the guy identified it in a millisecond.

About six hundred million years ago, complex eukaryotic cells with their efficient metabolisms started coming together, forming cooperative communities. Meet the Ediacarans, the first animals. We've found impressions of their soft bodies in rocks around the world. Most were small, but some were a meter long. They were fan-shaped, feather-shaped, worm-shaped, and jellyfish-shaped, but lacked the symmetry or structures of any modern animal group. They were ethereal and otherworldly, and they make a unique design for an easily identifiable phone case.[26]

Multicellularity evolved over and over again, independently, and in multiple different lineages.[27] A "collective existence" was so advantageous that natural selection gravitated repeatedly toward a body made up of many cells.[28] Larger bodies meant fewer things wanted to eat you, and in turn, allowed predators to evolve.

POND-SCUM-TO-PUDDY-TATS

Ancient Egyptians called the Nile River *Au* or *Aur*, a word that means "black." Every year, rainfall in the Ethiopian Highlands floods the Nile, and black silty mud overflows the riverbanks. For millennia, farmers depended on the floods to deposit rich soil, turning the desert into productive farmland.

Also arriving with the rich muddy soil were frogs, tons of them. The froggies never made an appearance during the dry season; they only arrived with the mud. The logical conclusion? Mud gives birth to frogs.

Frogs weren't the only life arising from nonlife in the premodern world. Sides of dead meat, hung outdoors in prerefrigeration days, gave birth to maggoty worms, of course.

Francesco Redi, a seventeenth-century physician, was dubious and conducted what is believed to be the first real science experiment. Redi's conclusion? Rotten meat doesn't make flies, only flies can make more flies. Two hundred years later, Louis Pasteur demonstrated that bacteria only arise from other bacteria, delivering the knockout blow to the idea of "spontaneous generation"—life from nonlife.

Did Redi and Pasteur likewise deliver the death blow to the modern theory of abiogenesis, the chemical evolution of life? Dr. Elizabeth Mitchell, writing for the creationist site Answers in Genesis, thinks so. Attempting to trace life back to chemical origins, says Mitchell, is nothing more than "willful ignorance" by scientists: "Molecules to man evolution remains as fictional and imaginative as ever."[29]

Abiogenesis is nothing like the archaic idea of spontaneous generation with full-grown frogs and flies emerging from nonliving materials. Abiogenesis occurred over hundreds of millions of years—first assembling the building blocks of life and eventually generating the first cells. Abiogenesis and spontaneous generation are not the same—at all—yet creationist organizations still equate the two.[30]

Elizabeth Mitchell admits that abiogenesis *might* be plausible if the earth is billions of years old. But Mitchell approaches all science evidence with the presupposition of a six-thousand-year-old earth—created fully formed in six days—so abiogenesis remains a fiction. Creative props to Mitchell, however, for dubbing the process "pond-scum-to-puddy-tats."[31]

We don't know, and may never know, the exact evolutionary pathway of the chemical beginnings of life. But that does not mean we know nothing.

Amusing analogies, impossible odds, big math, and laughable probabilities are straw-man arguments appealing to incredulity. But when we start with evidence, not presuppositions, a path to life is plausible.

DISCUSSION PROMPTS

1. Tornados in a junkyard and monkeys typing Shakespeare are two popular stories told by apologists in an attempt to "prove" the existence of God. Have you heard or read other apologetic stories or analogies told as evidence for God, intelligent design, or special creation? If so, briefly describe the point of the story.
2. Explain the meaning of this statement: Physics and chemistry place constraints on biology.
3. What characteristics of deep-sea thermal vents make them likely locations for the "nurseries of life"?
4. Famous biological anthropologist Eugenie Scott noted that there is a lot of "iffy stuff" in abiogenesis. We don't know, and may never know, each and every step in the process. Does this bother you? Why or why not?
5. Membrane-bound organelles were a landmark innovation in the story of evolution. Why were membrane-bound organelles such a game changer for early cells?
6. Multicellular organisms evolved many times, in many different lineages. What advantages does multicellularity provide?
7. Oxygen was poison to most of the earliest organisms. However, organisms able to adapt to an oxygenated world had an advantage over organisms that could not use oxygen. Describe the advantages of an oxygen-fueled metabolism.
8. Is abiogenesis simply a modern version of

spontaneous generation? Compare the science concept of abiogenesis to the ancient idea of spontaneous generation.

9. How does a presupposition of a young earth hinder acceptance of evidence for abiogenesis?
10. Life from nonlife is a challenging concept. What questions about abiogenesis remain unanswered for you? In your opinion, what evidence for abiogenesis (if any) is most convincing?

3

THE EXPLOSION THAT WASN'T

He was a Victorian gentleman, born to privilege. He was sent to university to become a doctor like his father, but he left medical school, traumatized after witnessing anesthesia-free surgery on a child. Seminary was also in the plan, but that, too, was a no-go.

His first love was beetles. He was mad about collecting them. So enthusiastic was he about one rare find that he popped it into his mouth to free up his hands for more. Beetle spews burning fluid, man spews beetle out of his mouth, beetle escapes scot-free.

He signed on to a round-the-world trip as an intellectual companion to the ship's captain. The trip was supposed to take two years, but in the spirit of Gilligan's extended "three-hour tour," the trip lasted five years.

The young Charles Darwin returned from his voyage a changed man. Darwin soon made a name for himself in the British scientific world with his exotic collections and his descriptions and drawings of animals, plants, fossils, and geologic formations from the coasts and islands of South America.

Ever the analyst, Darwin decided to marry after constructing a two-column pros-and-cons list for marriage, including this romantic point in the "pro" column: "constant companion and friend in old age . . . better than a dog."[1] He married his first cousin, Emma Wedgwood (from the pottery family), and theirs was a loving relationship with mutual respect and support. Darwin doted on his children and included them in his studies of bees, barnacles, carnivorous plants, and pigeons.

An older Darwin ambles about the grounds of his country home. Down House sits on acres of beautiful gardens, just right for a family man and natural scientist but close enough to London to participate in scientific society. I imagine him sadder now. In the years since his marriage, Charles and Emma lost three children, including their beloved eldest daughter, Annie.

In the decades following his famous voyage, Darwin wrote and spoke about the common ancestry of life and the evolution of species by natural selection, but only to trusted colleagues and friends. Darwin was internally conflicted, so he kept all his skeletons in the closet.

Oh, Darwin was not conflicted regarding the truth of his convictions. Darwin dreaded the fallout of his ideas in a world where all life appeared suddenly and complete in their current forms—at least, that's what respectable people believed. Even Darwin's wife, who supported his research and even edited his writings, feared for his soul. Darwin was finally convinced to publish when a younger colleague's similar conclusions about natural selection threatened to scoop Darwin's big idea.[2]

DARWIN'S DILEMMA

Regardless of the field of science, researchers are expected to delineate the limitations of their studies. Was the study constrained by time? Was the sample size smaller than they would have liked? Were there outside factors that may have influenced the findings? All legitimate, peer-reviewed research includes sections outlining weaknesses in the study and unanswered questions.

Charles Darwin's masterwork, *On the Origin of Species*, was published in 1859. It detailed decades of observations, collections, evidence, and analysis. It was truly a landmark work, and science has never been the same. Although we now know far more about evolution than Darwin ever knew, the heart of the theory—descent with modification and natural selection—remains the foundational principle of all modern biology.

Importantly, Darwin knew what he didn't know. He did not know, for example, how traits were passed from parents to offspring; he just knew that they were.

Darwin lived during the infancy of paleontology. Many Victorian geologists believed that the strata of Europe, along with its intrinsic fossils, were representative of Earth's entire geological history.[3] There were stellar nineteenth-century paleontologists, to be sure, but modern fossil evidence radically eclipses the knowledge of the Victorians.

When Darwin published, the oldest of the *known* geological strata was the Cambrian layer (540–485 million years ago). The icons of the Cambrian are the trilobites—a large family of roly-poly bug-looking creatures . . . if roly-poly bugs were sixteen inches long and hailed from the *Star Wars* universe.

Roly-poly bugs (which aren't bugs at all) are crustaceans (like crabs and shrimp), as were the trilobites. Although trilobites grew quite large, most were less than four inches long, and some were tiny, like a grain of sand.

And like their roly-poly distant cousins, they could roll up in a ball for protection.

These stars of the Cambrian had highly developed eyes, a digestive system, a hard exoskeleton in three parts, and a segmented middle. Trilobites populated oceans all over the planet.[4]

Here's a problem: trilobites appear suddenly and dramatically in the fossil record. And for Darwin, whose earth-shattering theory depended on slow and gradual change over time, it was a worrisome problem.

It wasn't just the hard-bodied trilobites. The Cambrian explosion, as it is often called, is the most intense burst of animal evolution ever observed. Cambrian animals include feathery worms, shelled creatures, and, for the first time, fast and fearful predators capable of catching and crushing prey. Also debuting in the Cambrian were early versions of many major animal groups alive today.

The Cambrian gave Darwin a headache. In his private notebooks, he described the gaps in the fossil record as "pages torn from a book."[5] The Cambrian animals were obviously complex, but they apparently had no ancestors. This paradox is known as "Darwin's dilemma."

Far from shying away from the problem, Darwin faced it head-on: "To the question as to why we do not find rich fossiliferous deposits belonging to the earliest periods prior to the Cambrian system, I can give no satisfactory answer." The Cambrian was a liability, and Darwin knew it. Darwin called the situation "inexplicable" and a "valid argument" against his theory of evolution.[6]

Darwin remained confident in his theory, however, and predicted the existence of Precambrian fossils, noting that "only a small portion of the world has been geologically explored."[7]

FROM THE LAND DOWN UNDER

He was still a young man, but Australian Reg Sprigg had already made a name for himself in the "down under" geological community. The year was 1946, and with the end of the war, Sprigg was able to leave his military assignment (searching for uranium) and resume his passion: physical geology. One day, while roaming the Ediacara Hills in South Australia, Sprigg found a fossil of a jellyfish-like animal—in a layer of rock *older* than the Cambrian!

Sprigg talked about his find that very year at the meeting of the Australian and New Zealand Association for the Advancement of Science, but no one was terribly excited. Life before the Cambrian was a hard sell.

Soon after, Sprigg led a small expedition back to the spot in the Ediacara Hills where he found his fossil. While the "billy boiled" for lunch, Sprigg found a dozen additional specimens.[8] By the end of the expedition, Sprigg and his group had collected more than thirty fossils: more jellyfish-like animals, but also circular-shaped animals and fossils of the now iconic *Dickinsonia*, an animal with gills like a mushroom.[9]

Over the next several years, Sprigg received lukewarm receptions at conferences and from journal editors. Muddying the waters were previous claims of Precambrian fossils that turned out to be nothing more than serpentine limestone.[10] The scientific community was not yet ready to acknowledge the existence of multicellular animal life, millions of years before the Cambrian.

Decades of research followed, with more discoveries of Precambrian fossils around the globe, from Africa to Europe to North America.[11]

In 2004, 58 years after Sprigg's first Precambrian fossil find and 145 years after Darwin's *On the Origin of Species* was published, the Ediacaran period was officially adopted by the International Commission on Stratigraphy. Darwin was right—the fossil record was incomplete, and multicellular life predated the Cambrian.

By the way, if you've ever heard that there's a secret conspiratorial cabal of scientists out to deceive the public and cover for each other, remember Aussie Reg Sprigg. Scientists love to prove each other wrong. Peer review is rigorous and real.

A SLOW-BURNING EXPLOSION

A bar of Ivory Soap, microwaved for one minute, swells and explodes into a puffy cloud many sizes larger than the original bar. Post-Easter, you can visit YouTube for creative ways to explode your leftover Peeps. And for a more serious look at do-it-yourself explosions, there's interesting science behind exploding grapes in a microwave.[12]

What did *not* explode, contrary to popular jargon, was the Cambrian.

We don't know who originally coined the phrase, but the concept of an explosive expansion of complex life is pervasive, from science writers to pseudoscience explanations for the beginning of life.[13] In the celebrated history of life on our planet, a twenty-to-forty-million-years span may seem brief, but it's hardly an explosion.

A more accurate description of the Cambrian is a "radiation" or a "slow fuse."[14]

The roots of the Cambrian explosion/radiation/slow fuse lie deep in the Precambrian. Those wacky Ediacarans included sponges, radial animals (like today's jellyfish), and bilateral animals (animals with a right and left side)—a wonderful diversity already present, prior to the Cambrian "explosion." The launch of modern animal groups was therefore not limited to the Cambrian but extended across the Precambrian-Cambrian boundary.[15]

Still, we find very few definitive Precambrian-to-Cambrian transitional fossils. Where are the missing links? Where are the ancestors of the trilobites with all their joint-legged, complex-eyes Cambrian star power?

The fossil record, like a family photo album, records the general trajectory of a family tree. A family album will be missing some months, even years. The album may be missing pages, or the photos might be blurry or smudged in some cases. Despite the wealth of fossils we have, fossilization is a rare occurrence. Imagine every animal, every plant, every bacterium to ever live (and die) on Earth, and you'll realize how rare fossilization is.

Is the "sudden" appearance of many new kinds of animals in the Cambrian due to some sort of miraculous and sudden explosive event . . . or due to an increase in the rate of fossilization? For the first 80 percent of life history, most life was single-celled, and all life had soft bodies. Although we find imprints of soft-bodied life, the fossil record is biased against soft bodies and favors animals with hard body parts, like shells, spines, plates, or bones. Rare species or short-lived species might not appear in the fossil record at all.

What, then, drove the change from the soft and squishy bodies of the Precambrian to Cambrian bodies with spines and plates? What drove the complexity of eyes, jointed legs, and predatory behaviors?

Likely, it was several drivers, working in tandem, compelling the burst of complexity and diversity seen in the Cambrian.

During the Carboniferous period (359 to 299 million years ago), atmospheric oxygen levels were around 35 percent, compared with 21 percent today. Darting about in the oxygen-rich air of the Carboniferous were dragonflies the size of seagulls, while eight-foot-plus millipedes scurried along the ground. Could an increase in Cambrian-era oxygen levels have thrown evolution into overdrive?

Oxygen-depleted seas in the modern world provide a clue. In areas where the seafloor is extremely oxygen-poor, only tiny worms live, feeding solely on microbes. In areas of the seafloor with slightly higher levels of oxygen, we see more diversity in animals, but the diet is still the same—microbes. But jump to a seafloor where oxygen is between 3 and 10 percent, and we find animals eating other animals. These are the predators—equipped to hunt, catch, and eat.[16]

AN ARMS RACE

Complexity is metabolically expensive. Animals use oxygen to extract energy from their food—did a modest rise in oxygen levels, just before the Cambrian, drive the evolution of complexity? Prior to the Cambrian, the formation and subsequent breakup of an ancient supercontinent led to massive erosion of land into the sea, increasing both atmospheric and marine oxygen levels.[17] Subsequent rising sea levels and freshwater niches provided calcium and phosphorus—raw materials needed to construct fossilize-able hard body parts.[18]

Animals not only adapt to their environments but they also adapt to each other. Imagine the advantage of even a simple eye—one that can distinguish light from dark or fuzzy shapes. You can see, you can hunt, you can catch and eat other animals. Imagine the advantage of a hard body covering to an animal *on* the menu. Imagine the advantage of jointed appendages in escaping from a predator. Predators and prey coevolve, locked in a one-upmanship arms race, driving diversity and complexity. A modest increase in Precambrian oxygen levels likely fueled the race.

We know that when an environment changes (relatively) quickly, species can evolve rapidly. Five million years ago, a brackish sea formed over southeastern Europe. Within a few tens of thousands of years (an eyeblink in the timeline of life), more than thirty new genera of cockles evolved in the new habitat. About this same time, a large freshwater lake, Lake Tanganyika, formed in East Africa. More than one hundred new species of cichlid fish evolved in this (relatively) short period of time.[19]

Rapid (relatively) environmental changes are also indicated in accelerated evolution during the Cambrian. Rising sea levels due to tectonic movement would create new shallow water habitats, open to species who could adapt. A brief time of glaciation in the Cambrian would lead to a loss of habitats—species either adapted quickly or they died out.[20]

Oxygen, environmental changes, and predator-prey coevolution are all credible explanations for the diverse radiation of life in the Cambrian. There's one more likely candidate: Legos. Not the plastic kind, but the kind made of DNA.

Hox genes are general-purpose control genes found in the genomes of all sorts of animals: flies, mice, worms, even humans. *Hox* genes are present in all bilateral animals (animals with a right and left side), and versions of *Hox* genes are found in animals with radial symmetry (pie-shaped animals like jellyfish). *Hox* genes reach deep into the animal family tree—at least 600 million years or more.[21]

Hox genes control the basic head-to-tail body plan in an animal: "put the head here!" and "put the tail down there!" and "here's where the legs go!" *Hox* genes are powerful; even small mutations can result in an extra pair of wings or even legs growing from the head of a fly.[22] Variations in bodies are the result of *Hox* genes turning on and off at different times and in different places during the development of an animal.

Multicellular animals need ways to coordinate trillions of cells. Control genes, including *Hox* genes, do just that. *Hox* genes are the flexible genetic Legos capable of fast-tracking evolution and driving the innovations we see in the Cambrian.[23]

FANNING THE FLAMES OF EXPLOSION

Not everyone is a fan of the explosion. Oh, those opposed have no problem with the Cambrian period itself—the issue is the big boom. The Geological Society of America is leading the charge to defuse the explosion and substitute more accurate descriptions like the "Cambrian radiation" or the "Cambrian diversification" or even the "Cambrian slow fuse."

Although we don't know exactly who first described the Cambrian as an abrupt all-at-once event, we do know who latched onto the idea with ferocity. As early as the 1960s, creationist tracts were popularizing the phrase.[24] And in the third decade of the twenty-first century, creationist publications persistently portray the Cambrian as a dramatic, all-of-a-sudden event.

Downplaying the Ediacaran and amplifying the suddenness of the Cambrian are favorite strategies in the creationist playbook. In fact, Jeffrey Tomkins of the Institute for Creation Research declares the Precambrian (the Ediacaran period) to be "devoid of invertebrates."[25] What's more, Cambrian fossils have "absolutely no evolutionary history," according to Jeff Miller of Apologetics Press.[26]

All those Ediacaran weirdos would like a word.

Even Charles Darwin is drawn into the fray. Poor Darwin—although he predicted the existence of Precambrian animals—is still held accountable for his ignorance of discoveries made decades after his death. Darwin's "dilemma"—the one Darwin himself acknowledged could torpedo his theory—has been solved, yet the argument is still made as if it never was.[27]

Sometimes the strategy is to overstate the diversity of the Cambrian animals.[28] Phillip Johnson is the architect of the "intelligent design" movement, a form of creationism embellished with science jargon and technical terms. In his best-selling book *Darwin on Trial*, Johnson argues for an extensive and modern-like Cambrian fossil record: "The single greatest problem which the fossil record poses for Darwinism is the 'Cambrian Explosion' of around 600 million years ago. Nearly all the animal phyla appear in rocks of this period, without a trace of the evolutionary ancestors that Darwinists require."[29]

While it's true that early versions of many modern animal groups are found in the Cambrian, also important is what we don't see: sea stars, crabs, insects, spiders, fish, reptiles, birds, or mammals. We don't even see land plants.[30]

While it's true that early versions of many modern animal groups are found in the Cambrian, also important is what we don't see: sea stars, crabs, insects, spiders, fish, reptiles, birds, or mammals.

Wedding the terms "Cambrian" and "explosion" supports a fundamental creationist article of faith: a recent and instantaneous appearance of life. Ignoring Precambrian fossils and portraying the Cambrian as comprehensive makes special creation plausible and discredits evolutionary change over eons of time.[31]

FOREGONE CONCLUSIONS

Three men, all with terminal degrees in engineering or earth science, took the stage at Liberty University for Convocation in March 2023. Liberty, a Christian university with an undergraduate enrollment of almost fifty thousand students, invited the men to present science evidence for the global flood of Genesis.

More important than the actual evidence, according to the speakers, is the *approach* to the evidence: "If one takes the Bible to be God-breathed and reliable, this sediment record logically must be the product of the flood judgment," said School of Engineering professor John Baumgardner. And furthermore, approached with the correct biblical lens, the evidence will always date the flood between four thousand and five thousand years ago.[32]

In other words, this is how we do scientific research: We begin with a foregone conclusion, then we look for evidence to back us up. And with a foregone conclusion, you'll never be wrong.

The Liberty professors are not alone. A frequent creationist explanation for the "sudden" appearance of fossils in the Cambrian is Noah's worldwide flood.[33] Of course, the presupposition is always "taking the Bible seriously," which in a creationist context means "taking Genesis literally."

The Cambrian mystery, says Answers in Genesis paleontologist Kurt Wise, is easily solved: "By starting our scientific investigation with a firm faith in the truth of the biblical account of the Flood, scientists can find a solution to the Cambrian Explosion."[34] But first, says Elizabeth Mitchell, we must "divest" the fossil record of a millions-of-years age. Then, and only then, can we understand that Noah's flood explains not only the Cambrian but also the entire fossil record.[35]

"Flood geology" finds its origins in the early twentieth century, inspired by visions of the Genesis creation as reported by Seventh-day Adventist founder Ellen G. White. The modern young-earth creationist movement, birthed in the 1970s, relied on flood geology to explain both the fossil record and geologic features like the Grand Canyon. According to flood geology, sea creatures would be the first to die. Subsequent fossil layers (with more complex life) represent the smarter and faster animals, those able to run to high ground as floodwaters rose. Humans, only found in the uppermost layers, were the smartest of all and were therefore the last to die.[36]

Were all flood-era humans speedy adults? Were there no infants or children left behind? Was no one elderly or sick or disabled? If you've witnessed devastating floods like those following Hurricane Katrina, you understand that violent floodwaters do not lay down organized sediments. Everything is a mishmash of dead plants and animals, not the predictable layers we find in the fossil record. And if Noah's flood explains the Cambrian, we are left with a puzzle: Why no sea stars, crabs, insects, spiders, fish, reptiles, birds, or mammals in the Cambrian layers?

Without a doubt, the Cambrian was an amazing chapter in life's history. How exciting to look back and see a sneak preview of life to come! But in their day, those weirdo Precambrian Ediacarans and those somewhat recognizable Cambrian animals were just who they were, animals adapted to the environments in which they lived. They weren't trying to be something they're not. They weren't bemoaning their transitional status—they were just surviving and passing on their genes to the next generation.

Early in the Cambrian we find some curious little creatures. They appear so early on, their roots likely dip down into the Precambrian. They had soft little wormish bodies, but fossil imprints reveal traits that are un-worm-like and vaguely familiar.

These were the first chordates—early versions of what would eventually lead to us.

DISCUSSION PROMPTS

1. Valid scientific studies are expected to delineate limitations of the study. Describe possible limitations of a modern research study.
2. Summarize "Darwin's dilemma." Describe the gaps in Darwin's knowledge as a nineteenth-century scientist.
3. Reg Sprigg is credited with the first discoveries of Ediacaran fossils. How did the process of scientific peer review delay official recognition of the Ediacaran period?
4. Paleontologists have jettisoned the term "explosion" to describe the Cambrian period and currently prefer terms like "radiation" or "slow fuse." Why did the Cambrian initially seem so "explosive"? Explain why "radiation" and "slow fuse" are more accurate descriptors.
5. Fossilization, in general, is rare. Why is this so? Why are Precambrian fossils so rare?
6. As atmospheric oxygen increased, animal complexity increased. Animals adapt to their environment, but animals also adapt to each other. Describe the evolutionary "arms race" during the Precambrian-Cambrian periods.
7. Not all genes code for proteins. *Hox* genes, a family of control genes, direct the construction of animal

bodies. What is the role of *Hox* genes in animal body building? How might control genes like *Hox* genes contribute to the vast and diverse array of animal life?

8. How do creationist organizations typically characterize the Cambrian? Why is a "sudden" Cambrian especially important for young-earth creationists?
9. Many creationists assume that fossils are the result of Noah's worldwide flood. Discuss problems with a "flood geology" explanation for the fossil record.
10. Creationists believe that science evidence should always be viewed through a "biblical lens." The modern scientific method starts with a question, then looks for answers in the evidence. Compare the "biblical lens" approach to evidence with the approach used in the scientific method.

4

A FISHY STORY

A bearded man stands at a podium on the deck of a large wooden ark, surrounded by pairs of animals. It's starting to rain, to really pour, and the man announces, "I'm sorry, we've overbooked. We are offering three hundred dollars to anyone willing to take a later ark." The *T. rex* gives a big thumbs up (as big as possible with his ridiculously undersized arms) and shouts, "We're in!" And then there's *The Far Side*'s dinosaurs (my favorite), sneaking a smoke behind the gym, with the caption: "The real reason dinosaurs became extinct."

In a game of "name an extinct animal," the first guess is always going to be "the dinosaurs." After that, most people would probably be stumped. In reality, more than 99 percent of all species ever to walk, swim, fly, crawl, and grow on our planet are extinct.[1] Everything living today—from plants to animals to mushrooms—is just a tiny drop in life's bucket.

Five times in Earth's history we see global calamities—five mass extinctions. The best known, of course, is the Cretaceous extinction sixty-six million years ago. A six-mile-wide asteroid strike in the Yucatan peninsula ended the age of the dinosaurs, along with the flying reptiles (pterosaurs), a variety of monstrous marine reptiles (mosasaurs and plesiosaurs), as well as the shelled ammonites and many plants.

The other four mass extinctions were not so sudden.[2] Most took place over tens of millions of years, a mere blink of a geological eye. Volcanic activity, global warming or cooling, and ocean acidification caused catastrophic environmental changes at various points in Earth's history.

The Ordovician-Silurian extinction (485 million years ago) killed 85 percent of species, including many of the trilobites. The Late Devonian extinction (383 million years ago) killed 75 percent of species, including the giant armored fish. The Triassic extinction (201 million years ago) killed 80 percent of species—including many reptiles—clearing an ecological path for the rise of dinosaurs.

The largest extinction of all is aptly called "The Great Dying." The End-Permian extinction (252 million years ago) killed 96 percent of all marine species (including the remaining trilobites), eliminated about three-fourths of all land species, and wiped out most of the world's forests.

Five massive dyings, yet here we are. Life finds a way.

It doesn't take a global catastrophe like an asteroid or a millions-of-years-long environmental collapse for extinctions to occur. Early on, Darwin understood that struggle and competition in nature favored the most "fit" for the environment at hand. Sometimes it was a localized natural disaster, sometimes it was disease, but most often it was environmental change of some sort. Sometimes change was in the form of competition—a species better adapted to hunting and catching prey (or better at hiding and avoiding predators) outcompeted other species.

As an environment changes, members of a population with adaptive traits survive; those without, die out. Over time, species become extinct, and others evolve to fill their niches.

IF IT AIN'T BROKE

Species don't "try" or "want" to adapt. The genetic toolkit for an adaptive trait must already be in place, somewhere in the population. If there are genes for an advantageous trait somewhere in a population, that trait will take off. Eventually, only individuals with the advantageous trait remain; the population has evolved. If there are no genes in a population providing fitness in a changing environment, bad news. If a body plan or a body covering or a behavior isn't working, a species must adapt or it becomes extinct.

But if a body plan or body covering or behavior *is* working, and continues to work throughout environmental changes, we end up with modern animals very much like the earliest forms of that species. Such animals have "basal" traits—traits like the animals at the "base" of the family tree. Here's a fact: Some species seem to never change.

Horseshoe crabs (who aren't actually crabs) have been around, virtually unchanged, for 540 million years. Contemporaries of the trilobites, horseshoe crabs managed to survive extinctions while the trilobites did not. The multichambered nautilus (dating back 500 million years) is easily recognizable in the fossil record. Coelacanths are stout lobe-finned fish, long thought to have perished with the dinosaurs, but in 1938 they were found living off the coast of South Africa.[3]

Here's the thing about evolution: Change is driven by environmental pressures. No pressure to change, no evolution. "Living fossils" give us a front-row seat to one of the foundational principles of evolution theory: survival of the fit enough. Not survival of the "fittest" or strongest or fastest or best at anything—just fit *enough* to survive and leave offspring in the environment at hand. If it ain't broke, don't fix it.

Modern animals with basal traits tell us about the earliest creatures in our family tree. Important to our conversation is a modern animal with roots deep in the Cambrian, a modern animal with many basal traits. He's not too impressive compared to the popular kids of the Cambrian, but his ancestors set a course for us.

THE DEBUT OF POO

What a difference a little poop can make.

Late in the Precambrian, we find the first animals with a right and left side, a front side and a back side, a head-end and a tail-end. Introducing the bilaterians (animals with bilateral symmetry), and, wow, did this little innovation take off. The addition of a head (with a mouth) gave these animals the ability to meet the world head-on, move around, and eat other animals—setting the stage for predators to come.[4]

We also see tube-like guts in the bilateral animals, with openings at each end. With a "through-gut," we have the debut of poo. Seldom celebrated, but the origin of feces likely changed the world.

When organic material is decomposed by bacteria, oxygen is consumed. But fecal pellets, formed in a through-gut, were dense enough to sink to the ocean floor before they decomposed. Oxygen levels in the upper ocean increased, opening new niches, just waiting to be filled via evolution.[5]

While still in the Precambrian, the bilateral animals diverged into two major lineages. Early in development, while the embryo is still a ball of cells, an opening forms. In one lineage, this opening becomes the mouth. These are the protostomes ("mouth first"), and for a brief time in their development, these animals are just a mouth. Protostomes include the worms, the insects, and the mollusks.

In the other lineage, the first opening becomes the anus. These are the deuterostomes ("mouth second"). Deuterostomes include the echinoderms (sea stars, sea urchins, sand dollars) and the chordates. Humans are chordates, so we are deuterostomes. At this point, I can't resist a *slightly* PG-13 biology joke: Every person you know, at one point in their life, was a *complete* (insert the euphemism of your choice)!

And another fun fact: Because humans are deuterostomes, we are more closely related to sea stars and sand dollars than we are to worms or insects.

OUR WORMY COUSIN

Since humans are chordates, we'll pick up the trail at this point. Fossil chordates are found early in the Cambrian and are recognizable by a rod-like structure running down the back.[6] It's not a backbone, however. About fifty million years later, animals with a backbone would arise from within the chordates, so hold that thought.

The lancelet is a modern chordate without a backbone. He's a tiny sand-burrowing marine animal, a wormish cousin of ours, many times removed. The lancelet is one of those "living fossils," very much like our first chordate ancestors.

The lancelet doesn't have a head, but there's definitely a family resemblance at the head end. Lancelets don't have a brain either, only a simple nerve cord running the length of their body, with a tiny swelling at the head end.

In humans, control genes map out the three basic regions of the brain: the forebrain, the midbrain, and the hindbrain. The lancelet has versions of these same genes, but instead of building a brain, the genes build the nerve cord and the tiny swelling at the head end. Amazingly, the recipe to build our complex brains first arose in chordates, over 500 million years ago.[7]

The lancelet, and a few other species of modern chordates, are invertebrates. They have no backbone, or any internal skeleton at all. Ninety-seven percent of all modern chordates are vertebrates, creatures like us, with backbones and internal skeletons.

Vertebrate or invertebrate, all chordates share four defining features: the notochord, a stiff rod of tissue running head to tail; a hollow nerve cord running along the animal's back; gill slits in the neck area; and a post-anal tail. In vertebrates, including humans, the notochord is replaced by the backbone and the nerve cord forms our spinal cord and brain during embryonic development.

Gills are retained by fish, but in other vertebrates like us, they disappear before birth. Most vertebrates have a tail, but humans and other great apes only have a tail briefly before birth. Brainless lancelet or Stephen Hawking—all chordates have these four defining features, placing us securely on the chordate branch of the family tree.

RISE OF THE VERTEBRATES

Hagfish definitely did not win the nomenclature lottery. Unfortunate name, but probably earned. The body of a hagfish is lined with slime glands and in less than a second, a half teaspoon of slime will expand to fill a large bucket. The slime is softer than Jell-O, looks like snot, and if you stick your hand in that slime-filled bucket, you'll pull your hand out coated with the stuff.

Hagfish remain virtually unchanged in their 300-million-year history. Although they are classified as vertebrates, hagfish have only rudimentary backbones made of cartilage. What they *do* have is a skull, which lands them at the base of the vertebrate branch of the family tree. Hagfish have no fins. They have a mouth but no jaws, just a gaping maw used to suck up bits of dead carcasses.

Fish were the first vertebrates, and hagfish are very much like the first fish. Although the jawless Slime Life works for the hagfish, 99 percent of modern vertebrates, including humans, have jaws, a backbone, and paired appendages of some sort. In the case of fish, the appendages are fins.

The first fish with backbones, jaws, and fins had skeletons made of cartilage, not bone. Modern sharks, skates, and rays retain this ancient trait. Ninety-five percent of fish, however, have skeletons made of bone.

Jaws were the real game changer. With a jaw hinged to the skull, vertebrates could mount a powerful bite. An animal who can bite is not dependent on simple filter feeding or sucking up bits of dead stuff. Not only could prey be grabbed, but it could be ripped to pieces for easier digestion. When an animal can bite and chew, a whole new buffet of food is on the menu. Many jawed fish, like the shark, evolved into apex predators and ruled the seas.

Although only 5 percent of all animals are vertebrates, the evolution of vertebrates was a watershed event in the history of life on our planet. Fish were the first animals with backbones and skulls—actually the first animals with bones of any kind. With their movable jaws, fish ate more and better food. With more and better food, they had the nutritional resources to evolve complex sensory organs and behaviors and bodies . . . complexity isn't cheap.

Five hundred million years after they first appear in the fossil record, vertebrates could walk, run, type, land on the moon, and compose Eras worth of music.

FISHY CONNECTIONS

Assembling for their thirtieth annual meeting, members of the British Association for the Advancement of Science were in a bit of an uproar. It was only a few months after Charles Darwin published *On the Origin of Species*, but Darwin wasn't there. Instead, notable Victorian men of science discussed/debated the book and Darwin's theory of evolution by natural selection in his absence.

Many spoke, but retrospective history focuses on only two debaters: biologist Thomas Huxley (later nicknamed "Darwin's bulldog") and the anti-evolution bishop of Oxford, Samuel Wilberforce.[8]

The truth is somewhere in the retelling, but with no recordings or live tweets of the event, that is all we've got. Wilberforce, eloquently and with authority, spoke in favor of a six-day biblical creation. As his parting shot (and expecting a laugh), Wilberforce asked Huxley: "Are you an ape on your grandmother's or grandfather's side?" Huxley didn't miss a beat: "I feel no shame with an ape as an ancestor; in fact; I'd rather an ape than a bishop who speaks of science when he knows nothing of it." Burn!

"You came from a monkey" wasn't quite the mic-drop moment Wilberforce thought it was. With modern paleontology, modern genetics, and a time machine, Huxley could have blown the bishop's mind: We aren't monkeys. We are actually fish. Remember clades from chapter 1? Once you are part of the fish clade, you can't evolve out of it. You can claim membership in other clades within the larger fish clade—the mammals, the primates, the humans—but once you're in the fish clade, you're in for always—it's the Hotel California.

Fish are as much a part of human history as are Neanderthals—the clues are as obvious as the nose on your face. Grab a mirror. See the little groove between your nose and your upper lip? That's the philtrum, and it's a connection to your fishy past.

Fish are as much a part of human history as are Neanderthals—the clues are as obvious as the nose on your face.

At one month of age, human embryos are virtually indistinguishable from other vertebrate embryos, including those of fish. Your eyes start out on the sides of your head, like a fish. Your top lip, jaw, and palate are part of the gill-like structures in your embryonic neck, structures common to all chordates. Your nostrils and the middle part of your lip are at the top of your head.

As a human embryo grows, these three areas come together like pieces of a puzzle. By ten weeks, if all has gone as it should, the face of the embryo has fused seamlessly. There will be no scar, only that little groove between the nose and upper lip. If the timing is off, the pieces won't come together as they should, and a baby may be born with a cleft lip or cleft palate or both.[9]

Then there's poor Charles Osborne. He hiccuped for sixty-eight years, unsurprisingly a Guinness World Record.[10] Osborne took his condition in stride, chalking it up to a hog-butchering accident in 1922. He should have blamed his fishy ancestors.

Breathing is controlled by the brain stem in fish. In fish, the nerves that *control* breathing and the gills that *do* the breathing are tucked in close together, surrounding the brain stem. The trip from the brain stem to the gills is short.

It gets complicated in mammals like us. We have the same brain-nerve setup as our fishy ancestors, but the setup is MacGyvered to accommodate our mammalian-style of breathing. We breathe using muscles in our chest and our diaphragm, both of which are a long hike from our brain stem. Nerves leave the brain stem and must travel all the way to the diaphragm, creating lots of opportunities for things to go wrong. Nerve spasms of the diaphragm (hiccups) are often the result.[11] A simple motor reflex inherited from our amphibian ancestors keeps the hiccups going.

Now, let's talk gonads. In male fish, the testes are high in the chest, tucked up near the heart, behind the liver. This arrangement, however, does not work in mammals, including humans. Sperm are fussy little Goldilocks cells, functioning only when things are just right. Too hot, and the little guys are malformed. Too cold, they die.

In humans (as in all mammals), the testes start out high up in the body cavity, just like in our fish ancestors. Before birth, testes descend through the abdomen, trekking to cooler space. As the testes push through the body wall, the scrotum forms as an out-pocket of the wall, holding the testes in a Goldilocks zone.

The trip from high up to outside leaves a weakness in the wall, and sometimes, part of the gut will protrude through the weak spot. Males are prone to inguinal hernias because of their fishy past.[12]

The body cavity works for ovaries, but human females can't completely escape their fishy past, gonad-ly speaking. Ovaries are located near the Fallopian tubes, but they aren't connected. There's a small space between ovary and tube. Ideally, an egg successfully crosses the gap and makes it into the tube. Ideally, a fertilized egg doesn't get turned around, go the wrong way, and exit the tube. Unfortunately, a fertilized egg doesn't always clear the distance between ovary and Fallopian tube and can implant outside the uterus. Ectopic pregnancies—any pregnancy outside the uterus—are usually fatal to the baby, and without intervention, often to the mother as well.

The explanation for this unfortunate gap is in our fishy history. Fish shed eggs directly from their ovaries to the outside world—no problem there. The Fallopian tube is an after-market add-on in mammals.[13]

LAND HO!

On a pleasantly warm day somewhere near the equator, a large fish was hunting in the shallows and mudflats of a freshwater floodplain. He poked his head up above the waterline, looked right and left, then dragged himself to another spot for a better view.

Three hundred seventy-five million years later, our freshwater floodplain is no longer close to the equator and rocks from its former existence lay frozen on an island north of the Arctic Circle. A group of paleontologists are digging for fossilized fish, residents of the freshwater floodplain of long ago. The team, led by fish paleontologist Neil Shubin, did not pick this frozen spot willy-nilly. The team wanted to dig in 375-million-year-old rock, and Ellesmere Island in the Canadian Arctic is one of a handful of places in the world with exposed rocks of just the right age.

Shubin and his team knew the age of rock where the earliest fish fossils are found. The team knew the age of rock where the first four-limbed land animals are found. The team predicted that a fish with traits indicating a "transition" to land would be found in rocks of an in-between age.

What they found was the fossil of a flat-headed fish, not the conical-shaped head typical of fish. Despite its crocodile-like head, it was definitely a fish—it had fins and scales. But this fish had more: inside the front fins were the familiar bones of an animal with four limbs: one big bone, then two bones, then smaller bones of a wrist, even the beginnings of fingers. And—it had a neck! "Most of the major joints of the fin are functional in this fish," noted Shubin. "The shoulder, elbow and even parts of the wrist are already there and working in ways similar to the earliest land-living animals."[14]

The fish was named *Tiktaalik,* which means "large freshwater fish" in the local Inuktitut language. *Tiktaalik* is considered one of the most important water-to-land transitional fossils we have. Still, it's important to understand that at the time *Tiktaalik* was hunting on a warm day 375 million years ago, he was just who he was. He was fit and adapted for his time, his niche.

The other fish found living with *Tiktaalik* were huge and vicious predators. Some were twice as big as *Tiktaalik*; some had teeth the size of railroad spikes. A fish with a neck like *Tiktaalik* could be on the lookout. A fish with a shoulder and a tetrapod-like limb could do a push-up and drag itself to safety. In a "fish-eat-fish" world, says Shubin, you get big, you get armor . . . or like *Tiktaalik,* you get out of the water.[15]

Curiously, Neil Shubin—a fish paleontologist—taught human anatomy at the University of Chicago Medical School. Shubin has expertise in both fish and human anatomy because of our "inner fish." Shubin's book, *Your Inner Fish,* as well as the PBS series by the same title, explore our fishy traits in detail. And for the littles, I love the picture book *Grandmother Fish,* but a kid's favorite adults will also find it charming.[16]

We have no way of knowing if *Tiktaalik* is a direct

ancestor of humans. But we do know that the first animal with limb and wrist bones like ours was a fish. The first animal with a shoulder and a neck like ours was a fish. *Tiktaalik* gives us a snapshot of the general trajectory from fish to land animals.

The first four-limbed animals—the tetrapods—were pioneers. They were the amphibious ancestors of modern frogs, salamanders, and newts. There was a whole other world beyond water's edge, and the tetrapods were here for it.

TINKERING WITH FISHYNESS

Is Genesis History? is a beautifully produced documentary with lots of on-site interviews at the Grand Canyon. The film, part of Focus on the Family's "Truth Project," features interviews with geologists and biologists, an engineer, a couple of philosophers, and an astronomer—thirteen people in all. The approach to science evidence for each man is the same: start with Genesis and make the science work.

There are the usual hyperbolic arguments against evolution, things like "you aren't going to get a shark to evolve into a bird."[17] Narrator Del Tackett can't believe that a loving God would do such an awful thing as evolution: "Would a God of love stand by and watch his creatures flopping around on the ground, trying to produce wings? Trying to produce lungs?"[18]

What most people who reject evolution know about evolution comes from anti-evolution resources. This very well-made science-y film portrays animals tragically "flopping" around and "trying" to evolve the things they need to live on land. This is an implausible scenario, and rightfully so. This is not the way evolution works.

Swim bladders and lungs are similar organs. Both are found in modern fish, but it's one or the other. Swim bladders are used for buoyancy, while lungs (in addition to gills) are used for breathing. Interestingly, some fish can use their swim bladders to breathe in a pinch.[19] Virtually all fish have an air sac—some use it as lungs, others as a swim bladder.[20] Studies using CT scans demonstrate that swim bladders and lungs are actually variations on the same organ.[21]

The genes that build swim bladders in fish are the same genes that build lungs in fish . . . *and* in humans. In fact, the genes used for air breathing and building limbs were present in fish fifty million years before the first tetrapod stepped out of the water. These genetic codes are still present in humans and in modern fish with basal traits.[22]

The conus arteriosus, a structure in our heart, facilitates efficient delivery of oxygen throughout the body. Modern fish with basal traits also have this structure. Bony fish do not, but they have the genes to build it.[23]

Synovial joints are freely moving joints, allowing for a broad range of motion in our wrists, knees, shoulders, ankles, and hips. The genes that build our synovial joints already existed in the ancestors of bony fish and are still present in modern fish with basal traits.[24]

The fishy gills all humans have as embryos become our larynx (voice box) as we develop.[25] The genes to build "brown fat," the heat-regulating tissue found in human babies, are also found in fish—*cold*-blooded animals who don't even *have* brown fat, for that matter.[26]

Contrary to Tackett's example of a poor pathetic creature tossed out onto land, writhing and gasping for breath, the toolkit to build lungs and limbs (and so many other things) was already present in fish. Lungs did not have to pop into existence. Evolution is a tinkerer. Existing traits are modified; existing genes are marshaled for new jobs. That's how evolution works.

Lungs before land. Limbs before land. Truly we are fish. Fish with feet.

DISCUSSION PROMPTS

1. Assess your knowledge of the planet's great extinction events. Did you know about the end-Cretaceous asteroid that wiped out the dinosaurs? Were you aware of the other major extinction events? Were you surprised to learn how close life came to complete obliteration?
2. Before reading this chapter, were you familiar with the term "living fossil"? Briefly explain what is meant by the term. Creationists often cite "living fossils" as evidence against evolution (life has changed very little since the creation week) or as evidence for a young earth (there hasn't been enough time for evolution). How would you respond to these arguments?
3. Why have "living fossils" like horseshoe crabs, coelacanths, and gingko trees changed so little over time?
4. With the evolution of a through-gut came the "debut of poo." How did the presence of fecal pellets drive evolution?

5. Were you surprised to learn that versions of the genes that build our complex brains are found in a brainless chordate dating back over 500 million years? What does this fact tell you about how evolution works?
6. Jaws were a landmark innovation in evolution. What advantages do jaws provide?
7. Neil Shubin is a fish paleontologist, but he taught human anatomy in a medical school. How did Shubin's expertise in fish anatomy prepare him to teach human anatomy?
8. The narrator in the film *Is Genesis History?* believes that evolution is incompatible with a God of love. What are your thoughts?
9. Genes that build swim bladders in fish are the genes that build lungs in humans. Were you surprised to learn that these genes existed around fifty million years before four-limbed animals stepped out onto land? What does this tell you about how evolution works?
10. "Evolution is a tinkerer." Explain.

5

WARM FUZZIES

Prince Albert, an educated and (for the day) modern man, was never content with his full-time gig of being husband and second fiddle to Her Majesty Queen Victoria, so he spearheaded the Great Exhibition of 1851. London's iconic glass and iron Crystal Palace, built specially for the event, housed all the marvels of the Victorian age and celebrated the Industrial Revolution.

In the park surrounding the Palace, a herd of life-sized dinosaur sculptures roamed long after the exposition closed. On New Year's Eve 1853, eleven guests dined inside the shell of an Iguanodon. Seated inside the head of the creature was Mr. Dinosaur himself: Sir Richard Owen, comparative anatomist, paleontologist, darling of the British scientific elite, and the man who originated the term "dinosaur."

Owen's expertise and influence were critical in the creation of the beautiful Natural History Museum in London. Despite his accolades, Owen was territorial regarding his position at the top of the British science establishment, and he did not appreciate the lavish attention given to another Victorian naturalist, a certain Mr. Charles Darwin.

Meanwhile, in a far-flung corner of the British Empire, a South African road builder was hauling out loads of dog-sized skulls with impressive canine teeth. The skulls were found in rocks 270 to 260 million years old and looked nothing like animals in the modern South African savanna.

As the go-to guy for all things anatomical, Owen was sent this load of curious fossils. The skulls were puzzling mosaics of reptile-looking heads and large mammal-like canine teeth. But to acknowledge the "in-between-ness" of these animals gave too much credence to Owen's rival and his upstart ideas of evolution and common ancestry.[1]

Owen stubbornly refused to consider that the South African skulls were mammalian ancestors and dubbed the fossils "mammal-like reptiles." Although the South African skulls had a single opening behind each eye socket (characteristic of mammals) and *not* the two holes characteristic of reptile skulls, Owen's inaccurate description persisted for more than a century.

OH, PIONEERS!

All credit to the original tetrapods, *Tiktaalik*'s cousins, who first ventured out of the water. They were the pioneers—the amphibious ancestors of today's frogs, salamanders, and newts. The world would never be the same.

But modern amphibians and their predecessors are tethered to water. Amphibians depend on water for fertilization—as with fish, males must swim over an egg cluster and release their sperm.

Water is not only needed for fertilization, but amphibian eggs must also incubate in water. Tiny and transparent, shell-less amphibian eggs (laid by the hundreds) would quickly dry out if exposed to air. Those pioneering amphibians could feed along the shores but could never venture too far afield.

Around 325–314 million years ago, we find evidence of a landmark event in evolutionary history, a game changer for four-limbed land dwellers.[2] There were some new tetrapods on the scene, and they were on the move.

Introducing the amniotes. Layers of membranes grew from the embryos of these animals, not a part of the embryo itself but originating from embryonic tissues. One membrane, called the amnion, was filled with fluid and surrounded the embryo with its own little personal pond. Another membrane filled up with nutritious, protein-rich yolk to feed the developing embryo. Surrounding it all was a shell, either leathery or hard and rigid, another protection against desiccation.

At last—an egg that could be laid on land. The amniotic egg was a costly enterprise, however. An egg with all these innovations required lots of energy and nutrients to produce.

Once a membrane-and-shell-covered egg was laid, it was impenetrable to sperm. Fertilization, therefore, had to be internal, before the membranes and shell were in place. This required cooperation between the male and female and the effort of finding and attracting a mate . . . the cost was adding up.

The costliness of the amniotic egg meant that the amniotes laid far fewer eggs than did their amphibian ancestors. Oh, but the benefits! No longer constrained to water's edge, the amniotes traveled far and wide, across the globe. A whole new world (literally) opened up. Niches previously off-limits provided new homes—in forests, deserts, mountains, trees, fields—and some amniotes even returned to the sea as air-breathing, four-limbed, aquatic creatures.

Eventually, some amniotes would forgo the shell and retain the embryo and membranes internally, giving birth to live young.

In a (relatively) short time of a few million years, the amniotes split into two distinct lineages. Maybe a population of amniotes was divided by a flood, or a storm, or some other catastrophic event—we'll never know. But once separated, each group adapted and evolved along its own path.

Whatever the cause of the division, one group evolved with two openings in the skull behind each eye socket. The other group evolved with a single opening behind each eye.

The group with two openings were the diapsids. The diapsids would eventually give rise to dinosaurs, flying pterosaurs, all the great sea reptiles like mosasaurs and ichthyosaurs, and to modern reptiles like snakes, lizards, crocodiles, and birds.[3]

Our interest is in the other group—the ones with the single opening behind each eye: the synapsids. Initially, the synapsids were hard to sort, at least from our modern perspective. They were definitely not reptiles, but neither were they mammals. It was this zoological no-man's-land that eventually led to mammals . . . and to us.

PERMIAN WORLD

If there had been any cartographers around 300 million years ago, creating a global map would have been a relatively easy task. There was only one enormous ocean, and instead of the familiar seven continents, there was only a singular giant: the mega-continent Pangea. Initially, Pangea drifted northward, creating a cool and humid climate in which amphibians thrived.

As the northern drift abated, the sheer size of Pangea meant that most of the continent did not get the benefit of ocean breezes and moisture. Pangea transitioned to a supercontinent of interior dry deserts. Amphibians who need wet swamps declined, and fortune favored the animals not tethered to water. Welcome to the Permian.

The Permian world is one of fern and horsetail forests, cone-bearing trees, and other plants you might recognize as modern. The warming climate favored plants that could colonize dry land. Millipedes two meters long crawl across the forest floor, and dragonflies with two-foot wingspans hover overhead. Three-foot-long amphibians with cartoonish boomerang-shaped heads scuttle about to find a damp spot.

Through a clearing stomps a ten-foot-long predator looking for his next meal. His stocky legs are splayed out under his body, crocodile-style. His most extraordinary feature is a five-foot-tall bony sail spanning the length of his back.

Meet *Dimetrodon*. His showy sail usually lands *Dimetrodon* in the dinosaur bin at the toy store, much to the consternation of die-hard dino-fans. But peek into his mouth (you wouldn't dare), and you'll find round incisors, big canines, and curved teeth along the cheek. These are not the teeth of a dinosaur or any other reptile.

Biologically, *Dimetrodon* and the other synapsids are "stem mammals," an extinct group of precursor mammals who evolved into modern mammals. As strange and dinosaur-y as *Dimetrodon* looks, it is much more closely related to you and me than it is to a dinosaur.

The folk music duo The Doubleclicks immortalized the quirky *Dimetrodon* in song—it's an ode to the misunderstood. Look it up. You'll be singing it all day:[4]

> When I am feeling this, one thing will make
> me smile
> There are other folks who never quite fit in.
> They're kinda mammals but they look like reptiles
> But I bet that difference never bothered them.

NOT YET MAMMALS

Dimetrodon was just one of several lineages of stem mammals evolving during the Permian. The variety of these definitely-not-reptiles-but-not-yet-mammal creatures was stunning. Some, like the wolf-sized gorgonopsians, sported huge saber-tooth canines. But unlike their saber-toothed tiger descendants, the gorgonopsians replaced their teeth on the regular, reptile-style. Others were huge head-butting herbivores with five-inch-thick skull bones.

These enigmatic stem mammals were the creatures Richard Owen misidentified as "mammal-like reptiles." They are mosaics of reptile and mammal traits, it's true, so we probably shouldn't be too hard on him. Still, Owen was primarily motivated by his jealousy of Darwin and his desire to stay within the good graces of the British elites, who, for the most part, believed that all animals were specially created and had not changed since the six-day creation week.

As the Permian drew to a close, more and more mammalian traits evolved, seen here and there, in these still quite reptilian-looking animals. Notable was their stance—gone were the sprawling crocodile-style limbs, replaced with vertical limbs. Some were even walking mammal-style on their toes. There are also clues that body temperature was rising, and teeth, jaws, and brains were all looking more mammalian.

Up next: the greatest catastrophe the world has seen to date. The End-Permian extinction is the closest life has ever come to annihilation. Massive volcanic eruptions spewed clouds of sulfur dioxide into the atmosphere. Carbon dioxide blanketed the planet and caused acid rain, ocean acidification, and global warming.[5] Ninety percent or more of all species disappeared.

THE SURVIVORS

As temperatures rose and the earth boiled, plant ecologies collapsed. In this world, large animals were disadvantaged. Most of the stem mammals did not survive the End-Permian extinction, but one group managed to weather the almost 100,000-year storm: the cynodonts.

The name means "dog tooth." The cynodonts had a mouth full of very mammal-like teeth: large canines and specialized premolars and molars, all capable of biting *and* chewing. Unspecialized reptile-style teeth could bite, but that's it. Reptiles usually swallow their food whole. Still, like reptiles, cynodont teeth fell out regularly and were replaced with new ones.

Cynodonts were smaller than their synapsid ancestors.

Mostly weasel-sized, they could burrow underground and shelter in place from bad weather, dust storms, and temperature swings. In a world of limited food supplies, small was an advantage. They didn't need as much food and could lie dormant if needed.[6]

The Permian period ended, the Triassic dawned, and the cynodont survivors emerged on the other side. Not only did the cynodonts survive, they flourished. What was their secret?

The story of *Gulliver's Travels* provides a hint. A phenomenon called the "Lilliput effect" describes the decrease in body size seen in animals following an extinction event.[7] The cynodonts, already on the small side, continued to shrink as the earth recovered.

Small animals (think mice and rats) reproduce with a big bang: they grow faster, mate earlier, have more babies, and die sooner than their larger counterparts. Within a million years of the Permian extinction, cynodont diversity recovered.[8] In a devastated ecosystem, widespread death creates opportunities for the survivors. The little cynodonts spread far and wide, filling niches across the planet.

WARM FUZZIES

Tiny was working for the cynodonts. In addition to the cynodonts, the Triassic world was also being populated by a group descended from the reptile branch of the amniotes: the dinosaurs. As the dinos went big, the premammal cynodonts went small. They were nightcrawlers, creeping out under cover of darkness to avoid being dino-stomped.

As the cynodonts shrank, they grew in their mammalness.

Have you ever wondered why modern birds and reptiles are often so brightly colored, while modern mammals usually wear sensible tones of gray, brown, tan, or black? Evolution works by selecting for traits that give an advantage. Traits that don't provide an advantage fade away. Tiny little burrow-dwelling, night-hunting creatures have little use for a keen sense of sight. During this time in our evolutionary history, stem mammals lost the sharp, color-sensing vision found in their reptilian cousins.[9] Most modern mammals can't see color. Humans can, but that's another chapter.

Though they were small, they were fierce . . . and fast! Reptiles, with their sprawling limbs, wiggle their spines sideways—think about the slither of a snake or a crocodile. Cynodonts walked upright, limbs situated underneath their bodies, with spines that moved up and down—think about the running gait of a cheetah or greyhound.

Not only did this new way of walking allow for more speed and agility, but it also facilitated greater lung expansion. Cynodont fossils lack ribs in the lower part of the torso, a telltale sign of a muscular sheet—the diaphragm—capable of drawing a load of air into the lungs.[10] By the end of the Triassic, cynodonts exhibited unmistakable evidence of the high metabolisms and the constantly warm bodies seen in modern mammals.[11]

Our little premammal mini-me could run faster and longer than the reptiles from which he diverged. The cynodonts were able to forage and consume loads of calories to fuel their warm, increasingly active, and athletic little bodies.

And here's a dramatic preview of coming events: Some cynodonts had tiny pits along the bones of their snout, just like the pits along the snout we see in modern whiskered mammals. The cynodonts might not have had keen sight, but with whiskers comes a keen sense of touch. It's also the first evidence we see of hair. And if the cynodonts had whiskers, they likely had body hair too—maybe sparse, but body hair, nonetheless.

THE BETTER TO HEAR YOU WITH

If we were to rewrite "Little Red Riding Hood" and set it in the unfolding story of mammal evolution, the dialogue would go something like this:

"What big jaws you have, Grandma!"

"The better to hear you with, my dear."

Reptiles cannot pick up sound vibrations from the air; they only sense ground sounds. To pick up ground sounds, reptiles lie down, then transfer the vibrations to a single bone in the middle ear. That high-pitched nasally song of the snake-charmer's flute is lost on the snake once it raises its head, cobra-style.[12]

Mammals, on the other hand, have extremely sensitive ears. They pick up sounds from the air and under water. Mammals hear high-pitched squeaks, low-pitched rumbles, even sound frequencies above human hearing. Mammals have ears like no other animal: three tiny bones in the middle ear amplify the sounds that reach the eardrum.

The story of mammals' exceptional hearing begins in an unusual place: in the jaw of a reptile. Reptiles have a large bone in their lower jaw that holds the teeth. Behind that bone are smaller bones, forming the jaw joint. Early stem mammals like *Dimetrodon* also had this jaw arrangement.

But then, the story takes a turn.

By the early Triassic, the lower jawbone carrying the teeth in stem mammals had grown larger, crowding and reducing the size of the already tiny bones of the jaw joint. As we go forward in time, the lower jaw continues to get larger and the joint bones get even smaller. By the time we see the most advanced cynodonts, the bone carrying the teeth had grown so large it made contact with the skull, creating a new jaw joint.[13]

And as efficient chewing evolved in mammals, so did big brains, fueled by all those extra calories.

So, what of the small bones of the former jaw joint? By the time the large bone grew to occupy the entire lower jaw, the smaller bones had been reduced to tiny splinters. Two of these tiny bones joined the single bone in the middle ear, creating the three bones of the mammalian ear.

And there you go: The bones that reptiles use to eat with became the bones we mammals use to hear with.

Keen hearing wasn't the only benefit of this new arrangement. A new jaw, with increasingly stronger jaw muscles, gave mammals a valuable party trick, one that is extremely rare among animals: chewing. Chewing allows digestion to begin in the mouth before food is passed on to the stomach. Chewing food allowed early mammals to take in more calories and do it with more efficiency.[14] And as efficient chewing evolved in mammals, so did big brains, fueled by all those extra calories.

In tandem with efficient chewing and big brains, we see the evolution of another uniquely mammalian trait: nursing the young.[15] Big brains are expensive, and fat-filled, high-calorie milk is just the thing to feed your growing (and toothless) babes.

By the end of the Triassic and on into the Jurassic, we find the first no-doubt-about-it mammals, the real McCoys, furballs with warm bodies, big brains, differentiated teeth, big jaws, and three bones in their middle ears. The largest were no bigger than a house cat, but most were tiny as shrews.

MAMMALS, UNCHALLENGED

The Mesozoic era, called the "age of the dinosaurs," lasted 186 million years—spanning the Triassic, the Jurassic, and the Cretaceous. Although the dinosaurs were the stars of the show, the tiny mammals flourished. As the mega-continent Pangea broke apart, our tiny little ancestors spread all over the globe. Early mammal fossils have been found in Great Britain, China, South Africa, and Greenland.

For 186 million years, the dinosaurs ruled, unchallenged. They grew to enormous sizes, the largest animals ever to live—to that point. Dinosaurs held the record until a descendant of the shrew-sized mammals, the blue whale, claimed the crown.

But first, disaster.

It was six miles wide, the size of Mount Everest, and weighed hundreds of billions of tons. It was traveling 80,000 kilometers per hour—hurtling toward Earth at a rate twenty times faster than a speeding bullet. At that speed, friction would heat the surrounding air to a temperature several times hotter than the sun.

On impact, the energy equaled 100 million nuclear bombs, exploding all at once. Within 1,000 kilometers of the impact, all life broiled. Literally. Sixty-six million years ago, an asteroid rained extraterrestrial death and disaster on the earth. It's been called "the day the Mesozoic died," and life was never the same.[16]

The Yucatan peninsula in Mexico is ground zero. Debris, soot, and vaporized rock blast into the atmosphere. Some of it orbits the earth, then rains back down. Much of it stays in the atmosphere, blocking out the sun. Wildfires rage, adding smoke to the air.

No sunlight, no photosynthesis. Plants die. Plant-eating animals die. Animals that eat plant-eaters die. Food chains collapse. More than 75 percent of species go extinct.

The mighty dinosaurs, rulers of the planet for 186 million years—gone. Along with the dinosaurs go enormous flying reptiles, the pterosaurs, and monstrous predatory reptiles in the ocean, the mosasaurs. In some parts of the world, more than half of plant species disappeared.

Survivors were small and lived in the water or burrowed underground. Winners included a lone branch of the dinosaur tree, the birds, and our tiny mammal ancestors.

In the Coral Bluffs of Colorado are one million years of exposed rock, just above the asteroid strike boundary.[17] Right above the strike boundary we find rocks packed with fungal spores. Fungus grows on dead stuff, and most of the post-dinosaur world was dead and rotting. Then we begin to see ferns. Ferns are often the first plants to return after a forest fire and are signs that plant life is recovering. Palms are next, a wealth of palms. A forest of palms. By 300,000 years after the asteroid, plant life rebounds.[18]

What of our little mammal ancestors? As plant life rebounded, so did mammals. One species had teeth suited for eating both meat and plants. By 700,000 years after the asteroid strike, we find larger mammals, one the size of a beaver.[19] Mammals were bulking up—but what was fueling the growth? We have clues.

As the world recovered, plants grew, spread, and diversified. And as new kinds of plants evolved, mammals specializing in those plants evolved—a classic example of *coevolution*.[20] No longer limited to a tiny size and an insect diet, mammals now had a whole new world of plants to choose from. Whole niches, previously dominated by eight-ton reptiles who could squash you with one step, were now open and filled with fresh vegetation.

Also found in Coral Bluffs are beautiful fossils of large bean pods (I've seen them), the oldest evidence we have of legumes. At 700,000 years after the asteroid, mammals were thriving, and protein-rich legumes were available on the menu. Taking advantage of the original paleo-diet, mammals grew larger than ever before.

FRANKENSTEIN'S MAMMAL

John Hunter was not the most famous dropout from the University of Edinburgh—that distinction belongs to the blood-phobic Charles Darwin, who decided that medical school was *not* for him. After a few decades in the Royal Navy, Hunter found himself a governor in one of Britain's new colonies: Australia. No longer able to dump its prisoners in America (thanks, Revolutionary War), the Brits instead shipped their unwanted residents to Australia.

Hunter was more interested in studying the unusual Australian plants and animals than he was in squelching the widespread corruption and illegal alcohol trafficking by prisoners and military officers. Hunter spent a day watching an Aboriginal hunter tracking some sort of bizarre critter in a lagoon. Hunter eventually captured one of the critters himself and sent it back to England, pickled most likely in some of that illegal alcohol.[21]

It was 1799, and English zoologist George Shaw thought he'd been punked.[22] It had fur like a mammal but no nipples. It had no teeth, but it had a beak. And it came with a most incredible story passed along by the Aboriginal people: The creature laid eggs. For the next eighty-five years, scientists debated the identity of this zoologic mishmash.[23]

MODERN MAMMALS

Only four groups of mammals survived the asteroid, and one of those groups went extinct thirty million years later. The three surviving groups gave rise to all modern mammals.

The mysterious Australian animal was a platypus. The platypus is a monotreme, one of the three lone surviving mammal groups. Monotreme fossils have only been found in Australia and South America, so monotremes must have evolved while Australia, South America, and Antarctica were still connected. Living monotremes are only found in Australia and nearby islands.

Just five monotreme species exist today—the platypus and four species of echidnas. Monotremes are the most basal of all mammals, meaning they retain many characteristics of their reptile-like ancestors. Like all mammals, they have hair, three middle ear bones, warm bodies, and they feed their young with milk. Unlike other mammals, monotremes do not have nipples—they secrete milk through the skin and their babies lick it from the mom's fur. And like their reptilian cousins, monotremes have a single opening for waste and reproduction.

But most jarring to those early naturalists who studied the platypus, monotremes lay leathery-shelled eggs. Monotreme hatchlings are called "puggles," which is probably the cutest word you'll read all day.

Marsupials, another modern mammal group, are commonly known as the "pouched" mammals—think kangaroos and koalas. Although the majority of modern marsupials are found in Australia and neighboring islands, the oldest marsupial fossils are found in the Americas. Like the monotremes, they migrated to Australia when continents were connected. Only the opossum represents team marsupial outside of Australia.

Marsupials have very undersized placentas, so babies are born early and underdeveloped. The hairless, limbless, honeybee-sized embryos must make the grueling climb to the mom's pouch and attach to a teat.

Most modern mammals are placental mammals, so-called due to their more complex placentas. These are the mammals we know best: our pets, our livestock, the exotic animals in our zoos and wildlife preserves.

A lineage of placental mammals eventually gave rise to the primates, and that's where the story of us continues.

GALLOPING GISH

The seventies and eighties were the heyday decades of "creation science" in the United States. Bills popped up in state assemblies across the country calling for equal treatment of evolution and creation science in public-school classrooms. Evolution-versus-creation debates were popular entertainment, especially with the anti-evolution crowd.

Duane Gish was a silver-tongued orator and the star of the creationist debate circuit. Gish participated in hundreds of debates, preferring a no-holds-barred approach with each speaker given forty-five minutes or more of uninterrupted time. Gish used his stage time to overwhelm his opponents with a barrage of questions and statements and straw men that could never be individually addressed and refuted, even within an hours-long time frame. The debating style known as the "Gish Gallop" is named for him.

Gish famously denied the existence of any transitional forms in the fossil record, mammal or otherwise.[24] Instead of addressing the evidence for the evolution of mammals, Gish preferred to joke about it all: "No one has explained yet . . . how the transitional form would have managed to chew while its jaw was unhinged and rearticulated or how it would hear while dragging two of its jaw bones up to its ear."[25]

Ignoring the evidence and receiving no answer to his straw-man questions, Gish declared victory for creationism and the foolishness of mammal evolution. Gish's arguments persist, and he is still a go-to guy for all the flagship creationist organizations.

Contemporary creationist writers like Elizabeth Mitchell make the same arguments made by Gish in the seventies and eighties: "Though no one has been able to demonstrate how these bones shrank to become useless for chewing, migrated from the jaw to the ear, and acquired a brand new and perfectly designed hearing function, this idea persists." Dr. Mitchell insists that the evidence for mammalian evolution is nothing more than an "evolutionary desire to 'connect the dots' on the tree of life."[26]

Here's the reality: The transition from the early synapsids to mammals is so well documented and so unbroken, it is difficult to determine the first "official" mammal. In fact, the synapsid-to-mammal transition is one of the best-documented and one of the most complete fossil records we have of a transition between two major groups of animals.[27]

We are now squarely in the age of mammals. Humans aren't far behind.

DISCUSSION PROMPTS

1. The amniotic egg was a game changer for land animals. Explain.
2. *Dimetrodon* is a favorite paleo-animal for many. Did you listen to the "Dimetrodon" song? Why is *Dimetrodon* not really a reptile and not really a mammal? Does this seem strange to you? Why or why not?
3. Cynodonts survived the end-Permian extinction. What adaptations aided their survival?
4. Paleontologists find it difficult to determine the first "official" mammal. Why is this so?

5. Which animal groups survived the asteroid extinction event that wiped out the dinosaurs? How were early mammals equipped to survive this disaster?
6. We see an increase in mammal size after the demise of the dinosaurs. Explain.
7. Duane Gish was a star debater, champion of creationism, and critic of evolution. His debate strategy was a barrage of questions followed by a rebuttal enumerating all the things his opponent failed to address. The "Gish Gallop" style of debate is his namesake. Have you ever heard a creation-versus-evolution debate? If so, what were your impressions? In your opinion, are debates a useful means of "hearing both sides"?
8. The jaw-bones-to-ear-bones transition in mammalian evolution is a point of contention for many creationists. "How did a transitional animal chew?" and "How did a transitional animal hear?" are often repeated arguments. Using what you know about mammalian evolution, how would you respond?
9. How does evolution explain the coloration of most modern mammals?
10. Monotremes like the platypus have many *basal* traits. What is meant by this term? Which traits in monotremes are considered basal?

6

PLANET OF THE APES

He wasn't just a little lost—he was *really* lost. He was also really big—twenty-three feet long. We have no way of knowing where he thought he was going, but we do know he never got there.

Our off-track traveler, a beaked whale, was busy navigating his home waters in the Indian Ocean when he took an unfortunate wrong turn off the eastern coast of Africa. He entered a deep-water river and swam inland for hundreds of miles. Our whale wasn't enormous by modern blue whale standards, but he was too big to make a U-turn in the river. Seventeen million years later, we found his fossilized skull in an ancient riverbed in what is now the Turkana region in Kenya.

It's not common, but whales are occasionally stranded in rivers. In 2006, a sixteen-foot-long northern bottlenose whale was found about 43 miles inland, stranded in the Thames River in central London.

Our Kenyan whale was found much farther inland—460 miles from the coast—a long trip for an ill-fated ending. Even so, the length of the trip is not what makes this story so incredible. Although whales sometimes travel long distances up rivers, they won't swim uphill—maybe 3 feet above sea level, at most.

The Kenyan whale was found at an elevation of 2,100 feet.

When our whale was stranded, East Africa was at sea level. It was a dense forest of tightly packed trees in a wet, humid climate.

East Africa sits on top of a hot spot of magma. The magma heats up the earth's crust, causing it to split down the middle like a pie when it bakes. Sometime between twenty million years ago and now, the earth's crust (like the pie) started splitting in northeastern Africa in a process called "rifting." Rifting produced a deep hanging valley a half mile above sea level with uplifted shoulders on either side, as much as two miles high.[1]

The impact of the rifting on climate was dramatic. The rift's uplift prevented moist air from the Indian Ocean from reaching East Africa. Over time, eastern Africa changed from a flat, wet, dense forest to an area of mountains, plateaus, grasslands, and desert scrub.[2]

But when in that twenty-million-year period did the shift occur? Our seventeen-million-year-old lost whale essentially timestamps it for us. Using the current elevation of the plateau where the whale was found and case studies of the steepest river grades, the elevation of the area at the time of the stranding can be calculated. Our whale was stranded at ground zero of East Africa's transition from dense forest to grassland savannas.[3]

In a dramatic line of demarcation, modern chimpanzees are found only west of the Rift Valley in the wet woodlands, and fossils of early human ancestors are found only in the grasslands of the east.[4] Apparently, environmental pressures favored primates who could stand up, straighten up, and walk long distances on two feet.[5]

About five or six million years ago, our unique history began. The East African Rift Valley is literally the cradle of our species. Our wrong-way whale reveals a key piece of the puzzle in our own story, the story of humans.

PURGATORIUS

The teeth were tiny—only three millimeters from root to crown. Blink, and you'd miss them. But paleobiologist Leigh Van Valen knew his post-dinosaur mammals, and he recognized that these teeth were noticeably different. The teeth belonged to a little guy who lived a mere hundred thousand-ish years after the last nonbird dinosaur perished.[6] The tiny mammals doing their best not to be dino-stomped during the reign of dinosaurs had tall, sharp teeth—the teeth of an insect eater. Van Valen's teeth, however, were the teeth of a fruit-eater. The new species was named *Purgatorius* after the hill where it was found in the Hell Creek region of Montana.

A collection of *Purgatorius* bones were eventually found, along with those of a closely related species. Stellar finds included ankle, shoulder, and hip bones, all with highly mobile joint surfaces: *Purgatorius* and his close kin were clinging to branches and climbing trees.[7] Shortly (geologically speaking) after the demise of the dinosaurs, early primates moved from the ground to the trees and were eating fruit found at the tips of branches.

Van Valen and the others who studied these fossils are certain: *Purgatorius* and his close kin are the earliest known primates on the tree of life.[8]

DIVERSIFICATION OF PRIMATES

By fifty-five-to-fifty-six million years ago, early primates had moved into North America, Europe, and Asia. Earth was a hothouse at the time, with blankets of CO_2 from volcanoes causing a 170,000-year-long heat wave. Many deep-sea species went extinct, but we see a frenzy of evolution on land.[9] In addition to early primates, we see diversification and spread of rabbit-sized early ancestors of cows, sheep, camels, pigs, and horses.[10]

Early primates found an escape from the heat in their tree-top homes. Primates have flat nails instead of claws, opposable thumbs and toes, and long fingers—all of which made climbing, grasping, and jumping from limb to limb a breeze. Branch-grabbing hands and feet evolved in tandem with teeth suited for fruit eating—another interesting example of coevolution.[11]

Primates were the brainiacs of the mammal class. Fueled by high-calorie fruits, primate brains grew larger. The visual region of the brain grew especially big, but the olfactory center grew smaller. Primates were trading sharp smelling for sharp vision.[12] These early primates had big, forward-facing bugged-out eyes. Many could see in technicolor. Bigger brains and bigger eyes crowded out the nose, so snouts got smaller. Big eyes and pug noses—our first primate ancestors were really cute!

Primates continued to spread and diversify like mad. They moved into Arabia and into Africa. Lemurs branched in one direction on the family tree, monkeys in another.

Somewhere between thirty-two and forty million years ago, groups of adventurous little monkeys (along with some rodent pals) rafted from Africa to South America, probably on giant mats of storm-blown vegetation and chunks of earth broken off from the coastline. Africa and South America were closer then, sea levels were lower, and the monkeys likely island-hopped along the way.[13] These accidental travelers were the ancestors of the modern New World monkeys: howler monkeys, spider monkeys, and capuchins.

ENTER THE APES

The heat wave eventually ended and the climate cooled. About thirty-four million years ago, all primates in North America and Europe vanished. But in the warm African rain forests, primates flourished. The monkeys who stayed behind in Africa were the ancestors of the modern Old World monkeys: baboons, macaques, proboscis monkeys, and mandrills.

Then, somewhere between twenty-five and thirty million years ago, a new primate lineage diverged from the Old World monkeys; enter the apes. Apes are larger than monkeys, have a larger brain-to-body ratio, and, most notably, do not have tails.[14]

Proconsul was an early ape who lived between seventeen and twenty-three million years ago. Despite his distinguished-sounding title, he was not named for an ancient Roman governor. His namesake was a chimpanzee in a late nineteenth-century London vaudeville show who wore clothes, smoked a pipe, and attended board meetings with his boss.[15]

Proconsul was a mosaic of monkey-like and apelike traits. He had the shoulders and elbows of a modern ape but the arm and hand bones of a monkey. We also see both ape and monkey features in his lower leg and foot. Likely, *Proconsul* could run on all fours as well as climb trees.[16] Despite a medley of ape and monkey traits, *Proconsul* was no jumbled-up "missing link." *Proconsul* was quite successful—he lived for millions of years. *Proconsul* was a transitional ape only from our perspective, looking back in time.

The tail-less apes eventually diversified into the orangutans and gibbons of Asia and the gorillas and chimpanzees of Africa. About six to seven million years ago, a branch of chimpanzee-like apes split; one lineage led to the modern chimpanzees and bonobos, and the other lineage led to modern humans. The great apes (gorillas, chimps, humans), as a group, are called hominids. The human subbranch of the great apes are the hominins.

THE MARCH OF PROGRESS

Henry Fairfield Osborn was born into a rich and aristocratic family. He made a name for himself as a paleontologist at the turn of the twentieth century. For almost three decades, Osborn was the president of the American Museum of Natural History in New York City.

Osborn fought to banish any references to ape-to-human ancestry in his museum, but not for the reason you probably think. Osborn's objection was not religious—his opposition was much darker. Osborn was a big fan of eugenics and fascism and Hitler, and the thought of all of humanity (including white aristocratic humanity) sharing ape ancestry was a no-go.[17]

Tucked away from the limelight of the museum was an immensely talented yet undervalued woman (imagine that!) named Helen Ziska. Ziska was a museum artist, tasked with creating drawings and exhibits for display. Her ink drawings and three-dimensional exhibits featuring human evolution via the ape lineage championed the science and covertly resisted Osborn's dark persuasions.[18]

Overworked and underpaid, Ziska was nonetheless recognized by noted primatologist William King Gregory, who used her work in his publications. Ziska died in obscurity, but years later, an artist used her drawings as inspiration to create one of the most iconic pieces of art of the twentieth century. In 1965, Time-Life published a volume in their Life Nature Library with a pull-out chart originally titled "The Road to Homo Sapiens" but is most often called "The March of Progress." You've seen it—a crouching chimp, walking on all fours at the far left, then progressing one in front of the other is a parade of ape-ish creatures, followed by a caveman-looking guy, and finally, a handsome, perfectly postured modern human at the front of the line. Fifteen stops in all.

Helen Ziska was constrained by the limited fossil record of the 1920s, yet she stood up for the science at hand. Importantly, she stood against the evils of eugenics in the name of science.

BECOMING HUMAN

Time-Life's misleading "march of progress" is taken as gospel by many and has fueled decades of "if we came from monkeys why are there still monkeys" argumentative throwdowns. The story of human evolution is not a get-in-line march to the pinnacle of all creation. Far from a simple line, human evolution is a tangled bush with twists and turns, overlapping branches, and dead ends. Some branches are so intertwined, they had children together.

As of now, we have at least twenty-one identified species in the human family tree. We have no way of knowing exactly who are direct ancestors (like grandmothers and great-grandmothers) and who are relatives (like aunts, uncles, and cousins). Although we are not descended from our aunts, uncles, and cousins, they give us a good understanding of the traits present in our family tree. All the species in the human family tree, both ancestors and relatives, are hominins.

In the six or seven million years after we last shared a common ancestor with chimpanzees, we see a trend of less chimp-like features and more uniquely human features.

Hominin fossils do not record each step in human evolution, but they do record the general trajectory. In the six or seven million years after we last shared a common ancestor with chimpanzees, we see a trend of less chimp-like features and more uniquely human features.

And again, it is not a single-file march. In the last six to seven million years, various human ancestor/relative species overlapped each other in time. We find species with many human traits living at the same time as species with fewer human traits.

Two of the earliest uniquely human traits are both found in the skull. The spine of a chimpanzee enters at the back of the skull, holding the head at an angle. Chimps occasionally walk upright, but their skeletons are built for walking on all fours. For a creature to stand up straight, its spine needs to enter at the bottom/center of the skull. Find where your dog's spine enters his skull and compare to where your spine enters your skull.

About seven million years ago, we find hominin skulls with a more "middle placement" of the spine under the skull, more like modern humans than like chimpanzees. By 4.4 million years ago, hominins had spine placement firmly at the bottom/center of the skull, holding the head tall and upright for bipedal walking.[19]

We also see a uniquely human trait in early hominin teeth. In the oldest hominin fossils, canine teeth (the pointed teeth in front) are noticeably smaller than the canines of chimpanzees. Male chimps like to show dominance or aggression by baring their canines. Small canines in male hominins may indicate more cooperative behaviors. The trend to smaller and smaller canines continued, and by 1.8 million years ago, hominin canines were short and blunt like ours.[20]

And while we're in the mouth, take a look inside your own. Notice how your teeth are arranged along a curve, a parabolic arch. This is a uniquely human arrangement. Chimps have a long U-shaped dental arch, more rectangular. In early hominins, we see an "in-between" tooth arch, with a trend to the human-like arrangement.[21]

The femur bone of hominins already looked human seven million years ago. In chimpanzees, the femur is a straight shot from hip to knee, oriented vertically at a ninety-degree angle to the ground. Chimps can walk upright, but they waddle when they walk. Bipedal walking in chimps is not feasible for any length of time or distance. In hominins, the femur forms an angle between hip and knee, bringing the knees close together. As a result, the feet fall directly below the body's center of gravity, making bipedal walking smooth and stable.[22]

The trend to efficient bipedal walking continued. By 4.1 million years ago, hominins had the strong knee joints we see in humans. Every time you take a step, you momentarily stand on one leg, stressing your leg bones. Strong knees help absorb the stress.[23] By 1.9 million years ago, we see the enlarged femur heads and long legs of a strider and runner.[24]

By 3.2 million years ago, the hominin pelvis was shorter and broader than in apes, and by 1.9 million years ago, the pelvis looked modern. A wide, basin-shaped pelvis is uniquely human. It is crucial in upright bipedal walking, but our pelvis also adapted to another uniquely human trait: birthing big-brained babies.[25]

Teeth, femurs, spines, and pelvises—all identify a species as hominin. And there's more: When we find a long, flexible thumb, long legs, low shoulders, S-shaped spines, and a flexible waist, we are looking at a fossil with uniquely human traits.[26]

BRAINIACS

The first human fossils we discovered in the eighteenth and early nineteenth centuries were big-brained hominins, not far removed (relatively speaking) in time from modern humans. It was reasonable to assume big brains came first in the story of human evolution. With a more complete fossil record, we now understand that long before we were making tools, developing language, and solving algebra problems, we were walking upright on two feet.

Big, complex brains are a uniquely human trait. The brain of a human adult constitutes 2 percent of the total body weight but consumes 20 percent of the body's energy supplies. But for the first four to five million years of human evolution, brains were small—bigger than a chimp brain, but not by much. Brain size grew slowly. We walked upright and made simple tools, but we weren't winning any Mensa awards.[27]

Brains don't fossilize, but skulls do. The brain sits snuggly inside the skull, so a cast of the skull cavity gives us a precise measurement of the braincase. Between two million and 800,000 years ago, bodies got bigger, and so did brains. The biggest jump in brain size, however, came between 800,000 to 200,000 years ago. In the time since humans and chimps last shared a great-great-great (many times over) grandmother, hominin brains have more than tripled in size.[28]

ELEMENTARY, DEAR WATSON

"There is nothing more deceptive than an obvious fact," said Sherlock Holmes.[29]

In the early twentieth century, England was basking in the glories of the closing Victorian age. The star of the age, second only to Victoria herself, was Charles Darwin. Paleontologists were making remarkable discoveries in support of Darwin's theory, and the hunt was on for evidence of human evolution, in particular. France and Germany already had bona fide fossils of early man, and accustomed to the limelight, England was feeling a bit left out at the (scientific) cool kids' table.

In 1912, amateur archaeologist Charles Dawson announced to the world the discovery of an ape-man near Piltdown in England. "Piltdown man" had the skull of a human and an apelike jaw. The find was heralded as the world's first "missing link" in human evolution.[30]

Red flags littered the ground, but the celebration carried on. The English scientific community was ecstatic: finally, the earliest human ancestor, and he was a Brit!

In the ensuing decades there were more red flags, more alarms. Turns out, Piltdown man was a cobbled together, doctored-up fraud—a human skull and the jaw of a modern ape, likely an orangutan. The teeth were filed down to make them look human and the bones were stained to make them look old.

Who masterminded the scam? Dawson was long dead—but did he act alone? Conspiracies abounded. Even Sir Arthur Conan Doyle, creator of the master-sleuth Sherlock Holmes, was a suspect. Some thought Doyle held a grudge against the scientific community, and with his medical background and his ability to create an intricate plot, many thought they had their man.[31]

Piltdown man is no more than an interesting story with the aura of a juicy conspiracy. It's a blip on the screen compared to the wealth of actual fossil and genetic evidence we have for human evolution. Yet, Piltdown man still populates the websites and publications and media of the biggest creationist organizations as "proof" of human evolution frauds perpetuated by modern scientists.

Creationist Randy J. Guliuzza believes modern scientists are just as deluded as those who fell for the

Piltdown hoax: "As evolutionists, these knighted but misguided scientists were rooted in the kind of imaginary extrapolationism that fosters visualizations and mental constructs that do not exist in a reality outside their minds. And their scientific prodigies are trapped in the same way."[32]

MAKING OF AN APE-MAN

My husband and his high school buddies were making faces and acting the fool outside the windows of a 1978 Best Catalog store, waiting for a girl in their youth group to get off work. Her boss, understandably annoyed, called the crew a bunch of "yard apes" and told her to go on.

Most people are OK with being a mammal, but an ape? No sir! And definitely not a yard ape.

Around six to seven million years ago, we shared a common ape ancestor with chimpanzees. That ancestor's lineage eventually split, with one branch leading to modern chimpanzees (still apes) and one branch leading to modern humans (still apes).

As the human apes evolved, we find species along the way who are mosaics, a mix of basal characteristics (like the chimp-ish common ancestor) and derived characteristics (unique to humans).

But can we trust the fossils? David Menton says no. According to Menton, Scripture is clear: God doesn't make "ape men." When scientists find an ape man, says Menton, one of three scenarios is at play: (1) the fossil is clearly ape but has been upscaled by overemphasizing vaguely human traits; (2) the fossil is clearly human but has been downscaled to seem primitive by overemphasizing ape traits; or (3) ape parts and human parts are cobbled together, often purposefully, to create an ape man.[33]

And who is the star example of scenario three? You guessed it—our old friend, Piltdown man. More than seventy years after the hoax was confirmed, Piltdown man is still the go-to guy to discredit human evolution. As I write, the creationist flagship Answers in Genesis features Menton's DVD *Three Ways to Make an Apeman* in its online store.

We don't have to upscale or downsize fossils to force them into the human lineage. In reality, we have an embarrassment of riches, fossil-wise. We have identified more than twenty species in the human branch of the family tree, and the transitions are ridiculously smooth.[34]

We have skulls with a chimp-like muzzle but a medium-sized braincase—both in the same cranium. We have hands that look human and an arm that is chimp-like, and these bones are found in articulation—still connected to each other, still fitted together—as they were in life. A bone can be long like a human's but thick like a chimp's. And it's not just parts and pieces—aspects of a bone, such as a femur head, can date an individual without the rest of the skeleton.[35]

The "must be all-chimp or all-human" camp of fossil analysis avoids an obvious question: Why would God specially create such perfectly intermediate species?

DISCUSSION PROMPTS

1. The earliest primates lived in trees. How do we know?
2. *Proconsul* was a mosaic of monkey and ape traits.

Proconsul was quite successful—he lived millions of years. What does *Proconsul* tell you about "transitional" species?

3. Are you familiar with the Time-Life "March of Progress" chart? In what ways is this chart misleading?
4. We have no way of knowing which ancient hominins are direct ancestors of *Homo sapiens* and which are close relatives in our human family tree. Regardless, what important things can our "relatives" tell us about human evolution?
5. Rarely do we find a complete early hominin skeleton. What features tell paleontologists that they are looking at hominin remains?
6. Despite being thoroughly debunked by scientists decades ago, Piltdown man is still a favorite of anti-evolution creationists. Why might this be so? What is the attraction to Piltdown man?
7. David Menton describes three ways to make an "ape man." What do Menton's "three ways" imply about the work of paleontologists and other evolution scientists?
8. We have fossil skulls with a chimp-like muzzle but a hominin-like braincase, all part of the same cranium. We have hands that look human connected to a chimp-like arm. Discuss: Would God create a species that is such an obvious intermediate?
9. Why did we initially believe big brains evolved before upright walking?
10. Most people are OK with being classified as a mammal but balk if classified as an ape. Why might that be so?

7

MEET THE FAM

It was one of those mall-based import stores—lots of bohemian hippie-chick clothing, incense, and multiple variations of Hindu and Buddhist art. One of those walk-through places that you don't necessarily plan to visit but you stroll through after your other mall errands to see if there's a tie-dyed skirt you *need*.

And there she was. At the back. Nothing special. No tags, no packaging, no other stock like her.

It was Lucy.

Well, not the original *Australopithecus afarensis* fossil discovered in East Africa in 1974, the one found by paleo-anthropologists while listening to the Beatles' "Lucy in the Sky with Diamonds" as they worked.

But a life-size Lucy replica. A museum-quality replica. Bones articulated and on a display stand that put me eye to eye with her. Eye to eye only because she was on a counter—Lucy stood only three and a half feet tall.

I nervously looked around. I felt like I had just discovered a treasure in plain sight. Is anyone else seeing what I'm seeing? I felt very Indiana Jones–ish.

There was no price on her, so, intrepidly I asked at

the register. I was torn between asking the price and not calling attention to my find.

Me: Is there a tag or packaging or brochure identifying this? Bored mall employee: Some kind of skeleton. It's been here a while.

Me: Ok. How much is it? Employee: It's on clearance. Nobody knows what it is, and nobody wants it.

Happy birthday to me. I just paid and walked out with her—no box, no packaging, no giant rolling boulder trying to crush me. I held her in front of me and carried her by her stand, through the mall, down the escalator. I lifted her arm and waved at a little kid as I passed him, but other than that, no one noticed as one of the most famous human ancestors exited the mall.

And that's why I have a life-size replica of the original Lucy in my home office.

FAMILY REUNION

Charles Darwin always believed human evolution began in Africa, yet there were no fossils to back him up. The few human fossils we had at the time were Neanderthals, but those were dismissed as diseased modern humans. A handful of human fossils were found in Europe and Asia over the next fifty years.

In 1924, Raymond Dart, a young anatomy professor in South Africa, was supposed to be dressing for a wedding. He was waylaid by a box of rocks sent to him from a limestone quarry. On top of the heap of rocks was a skull, obviously of an ape, but the brain was too big, bigger than that of an adult chimpanzee.

Dart could not contain his excitement, but he was the

best man, and duty called. Later, using his wife's knitting needles, Dart revealed the face, jaw, and teeth—it was a child. The opening for the spinal cord indicated that the child walked upright. Here was a human ancestor older than any to date, and it had a mix of human and ape traits.

Darwin vindicated! Not so fast. A decade before Dart's discovery, Piltdown man (we just can't shake him), the supposedly 500,000-year-old human ancestor, was getting all the press. Although red flags went up from the beginning, the fraud was not exposed for forty years. The Western scientific community *really* wanted a European origin for humans, so Dart's find, "the Taung child," was ignored for years.

Eventually, the 2.5-million-year-old Taung child (*Australopithecus africanus*) and other early hominin fossils placed the dawn of humanity decisively in Africa.[1]

We have identified at least twenty-one hominin species, to date. Don't be too beholden to a specific number—anthropologists will always debate here and there about whether a find is a new species or a variation in an already-known species. And importantly: This number doesn't represent the actual number of human fossils we have. Our current fossil holdings represent more than six thousand individuals.[2]

Let's hit the highlights, shall we?

The Earliest Hominin

We only know *Sahelanthropus* from his skull. At six to seven million years old, he's the oldest hominin fossil we have. That date puts *Sahelanthropus* right about the time we split from our last common ancestor with the chimpanzees.[3]

Sahelanthropus had the small brain of a chimp but a short face and small teeth like a human. The placement of the opening in his skull where the spinal cord entered indicates that *Sahelanthropus* walked upright. Was he a very early human ancestor? Or a side branch that died out? Regardless, *Sahelanthropus*'s small teeth and upright posture place him on the human side of the ape family tree.

Ardipithecus ramidus

We found Ardi in 1994. When she lived, 4.4 million years ago, she stood just under four feet tall. Ardi, and the others of her species (*Ardipithecus ramidus*), lived in eastern Africa. She was found with fossils of woodland creatures—pigs, monkeys, things like that. Apparently, Ardi lived in one of the remaining wooded areas of East Africa.

Ardi is definitely on the human side of the ape family tree but close enough to holler over to the branch where the chimpanzees were evolving. Like *Sahelanthropus*, Ardi and her kin had small, chimp-sized brains, but unlike chimps, they had small canine teeth and a centrally placed opening for the spinal cord.

It's the skeleton below her skull that's interesting. *Ardipithecus* had a shorter, bowlish-shaped pelvis—more human-like than ape, but her lower pelvis was more apelike. Ardi and her kin could walk upright, but probably not too fast, and probably not for long distances.

The feet of *Ardipithecus* add more to the story. Her

feet were flat, and her big toe was opposable, sticking out like a thumb. Ardi had feet more adapted to grasping tree limbs than to walking. Ardi's arms were long, but they weren't the arms of a chimp or gorilla—*Ardipithecus* wasn't swinging in the trees or hanging from branches. Likely, Ardi and her kin could climb in trees and move about the branches but could go to the ground and walk a bit when it suited them.[4]

Ardi defied expectations. Before we found Ardi, we knew—unequivocally—that humans and chimps were each other's closest living relatives. Before Ardi, we expected the most recent common ancestor to be very chimp-like.

But Ardi was not a knuckle walker like modern chimps, nor did she have the dagger-like canine teeth or the snout found in modern chimps. Ardi, so close in time to the common ancestor, was a surprise. We now understand that chimpanzees, like hominins, evolved unique traits once the two lineages split.[5]

Australopithecus afarensis

It was the early 1970s, so it's no surprise that the Beatles provided the background music at the campsite.

American anthropologist Donald Johanson and his graduate student were prospecting for fossils in the Afar region of Ethiopia. They were about to give up for the day when they spotted a hominin bone—an arm bone. The team found forty-seven bones in all, and all from the same individual. "Lucy in the Sky with Diamonds" was on loop as they worked back at camp because it was, after all, the '70s, and you played your tunes on a tape recorder. The hominin was eventually called *Australopithecus afarensis*, but she was known to the team, and to the world, as Lucy.

Lucy lived 3.2 million years ago. She was small, only three and a half feet tall, and was between twenty and thirty years old when she died. The team eventually found hundreds more bones representing many individuals and were able to fill in Lucy's missing pieces. Lucy and her kin are one of the best-known and best-represented species in the human family, with more than four hundred specimens.

Lucy and her species had a small brain and a long jaw like a chimp, but inside her mouth it was a different story. The canine teeth were small, and the dental arch was well on the way to a more human-like pattern. Her teeth were not arranged in the rectangular dental arcade of an ape, but they weren't quite the curved parabolic dental arcade of humans—they were somewhere in-between.

Lucy's arms were long, but not as long as those of a chimp. Her fingers were curved—did she spend time in the trees? Lucy's skull and trunk were a mix of human and ape traits, but below the waist she looked human. Her spine, pelvis, legs, and feet told an unmistakable story: Lucy and her species confidently walked upright on two feet.

Her femurs were angled inward toward her knees, a certain indicator of a bipedal stride. Her big toe was in line with the other toes, and she even had a slight arch in her foot.[6]

And if there was any doubt that Lucy's species was bipedal, Mary Leakey and her team put that doubt to rest, just two years later. At the Laetoli site in Tanzania, an eighty-foot-long trail of 3.6-million-year-old footprints was preserved in hardened volcanic ash. The prints were made by two individuals of Lucy's species, one bigger (probably a male) and one smaller (probably a female). The female's prints were deeper on one side, so she was likely carrying a baby on her hip, as mommas do. The prints are fossilized human behavior—a rare and valuable find.[7]

Lucy and her kin were not the only members of genus *Australopithecus*. We have identified several australopithecine species—some lived before Lucy, some after, some were contemporaneous. The Taung child was *Australopithecus africanus*, a species that overlapped Lucy's species and lived beyond them. For two million years, a potpourri of australopithecines lived in Africa, mostly in eastern regions.

The australopithecines were diverse, and over their two-million-year history, we see a trend to bigger brains. There was not a demarcated bigger-brain goal line that was crossed, but rather, from our perspective, we can identify a trend. We also see trends toward more humanlike teeth and skeletons.[8]

SIDE GROUPS

At around the two-million-years-ago mark, we find the first fossils identified as genus *Homo*, our genus. But hold that thought. At least three species of a genus called *Paranthropus* overlapped the last australopithecines and the earliest *Homo* groups in time.

The *Paranthropus* species were the thick-headed awkward relatives at the family reunion. They had small bodies, topped off with a broad face and flaring cheekbones. Their lower jaws were massive, and they had enormous molars. *Paranthropus* species had a bony ridge at the top of the skull to anchor their large chewing muscles. *Paranthropus* was built for bipedal walking, but their gait probably wasn't as refined as that of the australopithecines.

Paranthropus species are not ancestors of humans. They are an interesting hominin side group, evolving simultaneously, but are an evolutionary dead end.[9]

What we can learn from archaic hominins like Ardi and Lucy is the general trajectory of evolution from the last common ancestor of humans and chimpanzees to the first members of our genus, Homo.

So, who's the missing link? Where's the intelligent, speaking ape like *Planet of the Apes'* Caesar, who blurs the line between man and monkey? Was it Lucy? Was it Ardi? Was it one of their ardipithecine or australopithecine cousins?

Likely, we'll never know. What we can learn from archaic hominins like Ardi and Lucy is the general trajectory of evolution from the last common ancestor of humans and chimpanzees to the first members of our genus, *Homo*.

EARLY *HOMO*

Apologies for sounding like a broken record, but here's another reminder that human evolution was not a single-file march to the finish. It would be convenient if early humans fit nicely in a spreadsheet, with each group politely going extinct when the new group takes over. It's never that easy.

The earliest member of genus *Homo* lived between 2.3 and 1.5 million years ago. He's called *Homo habilis*, the "handy man." *Homo habilis* was first identified in Tanzania by the famous paleoanthropologist duo Louis and Mary Leakey. *Homo habilis* fossils have also been found at multiple sites in South and East Africa.

He's the "handy man" because he appears to be the first human ancestor to use tools. *Homo habilis* is associated with Oldowan tools, a type of simple cutting or chopping tool made by striking one stone against another to create a sharp edge.[10]

Homo habilis was similar in body and brain size to *Australopithecus* species, and many argued (and some still argue) that he belongs with the australopithecines. But the Leakeys were insistent—*Homo habilis*'s slightly larger brain and his association with tools promote him to the *Homo* genus. Over the years, more *Homo habilis* fossils were found. More complete skulls, hips, and legs secured his place within *Homo*.

Complicating matters are two other transitional hominins found in the same area. Are these two hominins simply variations on *Homo habilis*? Or are they so different they need their own species?

Naming a species is never an open-and-shut business. Some paleontologists are lumpers, some are splitters. Lumpers place all three in the *Homo habilis* column while noting the variations. In this case, however, most paleontologists are splitters, naming the variants as three separate species. *Homo rudolfensis* (with his more modern-looking skull) and the tall and slender *Homo ergaster* shared their East African home with *Homo habilis*, overlapping in time and space. These three early *Homo* ancestors had larger brains, smaller teeth, and flatter faces than all previous hominins in our family tree.[11]

And if you think three is a crowd, *Homo habilis, Homo rudolfensis*, and *Homo ergaster* also shared time and space with the more apelike *Paranthropus*. Who knows if they ever shared a campsite, but it was possible.

THE UPRIGHT MAN

A "family tree" is a useful analogy in understanding evolution, but a tree isn't the best botanical model for human evolution. Human evolution is more accurately a bush, a twisty, overlapping, tangled bush, and it will get bushier and more tangled before we leave this chapter.

In 1984, a fossilized skeleton was found in a dried-up riverbed in Kenya's Lake Turkana Basin. It was a boy, and judging from his teeth, he was between nine and eleven years old. He was a lanky kid, already 5'3" tall, with the long legs and the shorter arms of modern humans. "Turkana Boy," as he's called, had a slight curvature in his spine, probably scoliosis.[12]

Turkana Boy belongs to *Homo erectus*, the upright man. *Homo erectus* was successful—he survived for more than 1.5 million years. *Homo erectus* overlapped in time with *Homo habilis, Homo rudolfensis, Homo ergaster, Homo naledi*, a species of *Paranthropus*, as well as a late surviving australopithecine species.

Homo erectus was the first of our relatives to have human-like proportions. His longer legs and shorter arms indicate that he had left the trees for good. He was probably a runner, although not really the lean and athletic type. He was stockier than modern humans, with a larger and shorter ribcage. He had a flattened, chinless face with a prominent bony ridge across his brow.

Homo erectus was brainier than previous hominins, but still not as smart as modern humans. The *Homo erectus* brain was about half the size of ours, sometimes a bit more than half. Still, *Homo erectus* is associated with complex tools, like two-sided hand axes. With better tools and probable use of fire, he likely added protein-rich meat to his diet, fueling the growth of his bigger brain.

Homo erectus was bigger, smarter, and more dexterous than any other hominin to that point in time, and he did something no other human ancestor had ever done before: *Homo erectus* was a traveling man.

Homo erectus was the first hominin to leave Africa. His fossils have been found in the Middle East, Asia, and Indonesia. *Homo erectus* fossils in western Asia are between 1.8 and 1.85 million years old. One group of *Homo erectus*, known as Peking man, lived in caves outside Beijing, China, 750,000 years ago. Other adventurous groups of *Homo erectus* fashioned watercraft of some sort and made it to Java, where they lived until 250,000 years ago.[13]

THE HOBBITS

There is an interesting hominin group, *Homo floresiensis*, found only on the island of Flores in Indonesia. They lived between 100,000 and 50,000 years ago. They were tiny—only three and half feet tall; you'll often hear them called "hobbits." They had small brains, large teeth, no chin, and large flat feet (quite hobbit-y). They used tools and hunted small elephants.

"Island dwarfism" occurs when species are isolated for a long period of time on an island with limited food resources and few predators. It is likely that *Homo floresiensis* adapted to an isolated island life by getting smaller. The (now extinct) elephants hunted by the hobbits were also tiny—the smallest ever known.[14]

Fun fact for Tolkien fans: Modern humans overlapped in time with these real-life hobbits. As evolutionary biologist Jerry A. Coyne said, what's not charming about tiny humans hunting tiny elephants with miniature spears?[15]

The origin of the hobbits is a mystery. Are they descendants of *Homo erectus*? Or are they descended from another *Homo* species we've yet to discover?

HOMO HEIDELBERGENSIS

Meanwhile, back in Africa, another early human ancestor emerged—and he also had the travel bug.

More than a century ago, a human jaw was discovered in a quarry near Heidelberg, Germany. The teeth were small and human-like, but the bones were large and heavy. He's called *Homo heidelbergensis*, after the area where he was first discovered. Over the years, many more fossils were found across Europe—in England, France, and Spain—and possibly as far east as India.

The oldest members of *heidelbergensis*, however, are African: from the Lake Turkana region in Kenya, Kabwe in Zambia, and Elandsfontein, South Africa. Like *Homo erectus* before him, groups of *Homo heidelbergensis* set out from their land of origin to new territories in the north. *Homo heidelbergensis* lived between 300,000 and 600,000 years ago, and maybe as much as 800,000 years ago (depending on if you're a lumper or a splitter).

Homo heidelbergensis had the largest brain yet. Body proportions were similar to ours, but they were shorter and wider, with thicker bones. They had a flat, chinless face and a prominent brow ridge. Their teeth were arranged in a human-like parabolic arch but were bigger in size than those of modern humans.

Homo heidelbergensis was a toolmaker. He made stone hand axes and spears with stone spearheads. He was the first to control fire, build hearths, construct simple shelters, and hunt large game. We don't have definitive evidence, but he may have been the first to use animal skins as clothing. He made his home in very cold regions, so he was probably covering himself somehow.[16]

We've come a long way on the journey to us—at least five million years. Hang in there, things are about to look familiar.

MEET THE COUSINS

Deep inside a cave in Atapuerca, northern Spain, there is an abrupt vertical shaft, dropping down almost forty-five feet. At the bottom of the shaft is a thirty-foot-long chamber. The scene inside must have been shocking—a jumble of seven thousand hominin bones, belonging to at least twenty-eight individuals. There was a single quartzite hand axe in the mix, as well as the bones of a few wolves, foxes, lions, and lynx.

Sima de los Huesos ("pit of the bones") is 430,000 years old. Most of those in the pit were adolescents or young adults. Was Sima de los Huesos an ancient burial site? Were these ancient hominins interring their dead? Was the hand axe carefully chosen to accompany one of the deceased? Or was it a ritual of some sort? All are possibilities.

Initially, the remains were identified as *Homo heidelbergensis*, but something just wasn't right. These hominins looked similar to the most famous early human—Neanderthal man—although they lacked many classic Neanderthal features. Eventually, DNA testing provided the answer: The individuals in the pit were very early Neanderthals, *Homo neanderthalensis*.[17]

While some *Homo heidelbergensis* stayed in Africa, others ventured north. Neanderthals are descendants of these *heidelbergensis* emigrants. While modern humans were evolving in Africa, Neanderthals were evolving in Europe and Asia, between 400,000 and 40,000 years ago. Neanderthals lived all over Europe, as far west as Portugal and Wales. Neanderthals also lived in central Asia and in Siberia, and as far south as Israel.

Of all the human ancestors, you are most likely to recognize a portrait of a Neanderthal. Their big and wide nose, their double-arched brow ridge, and their short, stocky bodies are the stereotypical image of a "caveman." Neanderthals had a large, long skull, unlike the rounded globe-shaped skull of modern humans. Neanderthal brains were larger than those of modern humans but were proportional to the size of their bodies. They likely could communicate vocally, but we don't know if they had language. Neanderthals had a version of the FOXP2 gene, a key gene in human language.[18]

Neanderthals ate a lot of meat. Unlike their African cousins, they primarily lived in cold climates where plants would not be a consistent source of food. Neanderthals hunted big game like mammoths, bison, and reindeer with spears at close range. Fractures in Neanderthal bones are consistent with the fractures often seen in rodeo cowboys—guys who also wrangle big animals at close range.[19]

Neanderthals wore skins for clothing, they controlled fire, and they built shelters. Neanderthals decorated themselves with pigments and beaded ornaments. Neanderthals deliberately buried their dead and sometimes marked the grave with flowers—no other early human ancestor did that.[20]

A Neanderthal skull, 146,000 years old, tells an incredible story. The skull belonged to a child, a child who without a doubt had developmental delays and a host of other physical disabilities. In the hardscrabble life of a Neanderthal, she would not have survived infancy without the help of her social group. But here she was—six years old—a *six-year-old* girl with Down syndrome.[21]

We need to rethink the brutish, caveman stereotype.

There is another close cousin we need to meet. This cousin, known only from bone fragments, is the first member of *Homo* to be identified by DNA alone. The Denisovans, named for the cave in Siberia where they were first identified, were originally a mystery. The DNA in the bones was definitely *Homo* but was neither *Homo sapiens* nor *Homo neanderthalensis*.

Denisovans split off from the Neanderthals *after* the Neanderthals split from their last common ancestor with modern humans. The Neanderthals and the Denisovans are not our direct ancestors (like a grandmother) but are our closest cousins.

Denisovan bones, to date, have only been found in Siberia and the Tibetan plain, but that doesn't tell the whole story.

THE MUDDLE IN THE MIDDLE

Modern humans did not debut on the world stage with a triumphant "Hooray! We're here!" Our arrival was far more ambiguous. In fact, the emergence of anatomically modern humans is often called "the muddle in the middle."

Groups of *Homo heidelbergensis* left Africa and moved north, and from these immigrants evolved the Neanderthals and Denisovans. The best evidence (genetic and fossils) points to the African populations of *Homo heidelbergensis* as the most likely group from which modern humans emerged.[22]

It was not by any means a straightforward march to us. Modern humans evolved piecemeal, over 300,000 years and in many different locations. Archaic *Homo sapiens* lived all over Africa, geographically separated by rivers and valleys and cultural distinctions. Intermittent migrations brought tribal groups together, exchanging culture, language, tools, and, importantly, genes. Rick Potts, director of the Smithsonian's Human Origins Program, credits these exchanges with creating the human genome: "Everyone is African," says Potts, "and yet not from any one part of Africa."[23]

The oldest fossils of *Homo sapiens* are mosaics of modern and ancient features. For example, a 300,000-year-old skull found in Morocco has a modern face (flat and small), but the elongated braincase of older hominins. For 200,000 years, *Homo sapiens* mixed and matched:

chin/no chin, oversized brow ridges/small ridges, big globe-shaped braincase/elongated braincase. Sometime around 100,000 years ago, the distinct modern human body plan stabilized: slim and tall, with a giant brain in a globular cranium, a small face, pointy chin, and small brow ridges above each eye.[24]

Between 100,000 and 130,000 years ago, a few modern humans ventured north. Bones and tools have been found in Israel and as far away as China. These early forays out of Africa were evolutionary dead ends, however. The first small groups to leave Africa died out and did not contribute to the modern human genome.

The first large-scale migration out of Africa occurred around 50,000 to 60,000 years ago. Initially, the groups stayed close by in areas of Iran, Iraq, and Saudi Arabia.[25] Humans eventually spread into Asia and Europe and beyond, colonizing the planet from Australia to the Americas.

We aren't exactly sure how many people left Africa in the great migration—estimates range from 1,000 to 50,000.[26] Here is what we know: The founding group of migrants was relatively small. There is more genetic diversity in modern Africans than in the rest of the world's population combined.[27]

KISSING COUSINS

When *Homo sapiens* moved into Asia and Europe, we met some familiar faces—our long-lost cousins, the Neanderthals and Denisovans. But modern humans, it seems, were the last to arrive at the family reunion.

An extraordinary discovery tells us that Neanderthals and Denisovans not only met and interacted, but they also had children together. The cave in Siberia where Denisovan bones were initially discovered yielded an astonishing find: a bone from a thirteen-year-old girl with *equal* amounts of Denisovan and Neanderthal DNA. Half of her chromosomes were Neanderthal, half were Denisovan. We know she was Neanderthal on her mom's side because her mitochondrial DNA was Neanderthal.

It was as if we had a front-row seat. "Denny" (as she was called) was a first-generation offspring of a Neanderthal mom and a Denisovan dad.

We have yet to find evidence of a first-generation offspring of modern humans and Neanderthals or Denisovans, but we know they existed. We also know that the meet-ups occurred after humans left Africa—people with European or Asian ancestry have about 2 percent Neanderthal DNA, while modern African populations have almost none.[28] I've spit in a test tube, and unsurprisingly, 2 percent of my DNA is Neanderthal. All told, about half the Neanderthal genome is still around, scattered in bits and pieces throughout modern humans.

Human skin color, hair color, type 2 diabetes, increased fertility—all are linked to Neanderthal gene variants. Neanderthal genes are also linked to autoimmune disorders like Graves' disease and rheumatoid arthritis and to a risk factor for severe COVID-19. When *Homo sapiens* moved north, they had no exposure to diseases outside of Africa. Neanderthals, on the other hand, had 400,000 years to build immunity. Mating with Neanderthals provided a quick fix, but we were left with an immune system that is often overly sensitive.[29]

Denisovan DNA is found in modern humans but is far less common. Populations in Papua New Guinea and

Oceania have 4 to 6 percent Denisovan DNA.[30] Denisovan DNA is also found in populations living in the thin air of the Tibetan plateau. A genetic mutation found in both Denisovans and Tibetans is associated with adaptation to a low-oxygen environment.[31]

By 25,000 years ago, *Homo sapiens* was the last man standing. Neanderthals and Denisovans were extinct. Did we outcompete our cousins with our big brains, dexterous hands, and bodies built for long-distance running? Did the germs we carried with us from Africa defeat our northern cousins? Did we simply absorb our cousins into our species by our larger numbers and interbreeding? Most likely, it was a combination of all the above.

OUT OF THE MUDDLE

Who was the first speaker of Modern English? Who was the last speaker of Old English? We can identify a time when Modern English was the norm, but we can never identify the very first speaker. We can identify a time when Old English was the norm, but we can never pinpoint the last speaker. Languages evolve and move through periods of montage.

There was no "first" human. Try not to get bogged down in evolutionary games of who led to whom. In the journey to us, we simply don't know definitive ancestor-descendant lineages. As we get closer to modern humans and have the advantage of extracting DNA from younger (relatively speaking) bones, we can speak with more confidence.

What we have is a picture of the general trajectory of human evolution. We evolved as a population. We emerged from the muddle.

It's hard to not call our early *Homo* ancestors "human." I don't know. You don't know. We know they were not anatomically modern. We know that, cognitively, they weren't up to modern human brain standards. Still, we see echoes of ourselves in them. Even in the earliest ancestors, those before *Homo*, we see glimpses of familiarity.

In 1842, Queen Victoria's newest subject was about to be introduced to her sovereign. Lady Jane, called Jenny, was wearing her best dress and was sipping a cup of tea, which she spooned and sipped in a most lady-like way. Still, when Jenny was introduced to the queen, Victoria was repulsed, exclaiming: "how frightful and painfully and disagreeably human!"[32]

Jenny, in her new home at the London Zoo, was nonplussed. To Victoria, Jenny the young orangutan looked disturbingly human, dressed in human clothing (as the Victorians loved to do).

The implied relationship between apes and humans rattled many Victorians, so much so that Charles Darwin only vaguely referenced human evolution in his first book. A braver Darwin addressed the human question in his second book, *The Descent of Man*. Despite all his exalted powers, said Darwin, "man still bears in his bodily frame the indelible stamp of his lowly origin."

Indeed we do.

DISCUSSION PROMPTS

1. The "Taung child" was ignored for years. Why was such an important find initially given so little attention?

2. Which features of Ardi are chimp-like? Which features of Ardi are human-like? Did this combo of human and chimp traits make life difficult for Ardi? Why or why not?
3. Lucy walked upright on two feet. Describe the evidence indicating upright, bipedal walking by Lucy and her kin.
4. What important things about human evolution can we learn from early species like Lucy and Ardi?
5. Naming a new hominin species is not always a straightforward exercise. Why is the process often complicated?
6. Early human ancestors overlapped each other in time and, perhaps, even in space. Does this surprise you? Does this change your perception of human evolution? If so, how?
7. Neanderthals are stereotypically portrayed as brutish cavemen. What do we know about the culture of Neanderthals? Does this change your mental image of Neanderthals? If so, how?
8. The arrival of modern humans was an ambiguous event. Explain this phrase: "modern humans evolved from the muddle in the middle."
9. Were you surprised to learn how much Neanderthal and Denisovan DNA is found in modern humans? How do you feel about that fact?
10. In his book about human evolution, Charles Darwin observed, "Man still bears in his bodily frame the indelible stamp of his lowly origin." What kinds of emotions does this statement evoke for you?

8

IT'S IN THE CODE

It was early spring, 1986. Ruth Coker Burks, a twenty-six-year-old single mom, was visiting a friend in a Little Rock hospital. She watched curiously as three nurses drew straws in the nurses station. The nurse with the short end obstinately proclaimed, "I'm not going in there."

"There" was a room at the end of the hall. The door was covered with a tarp and a biohazard sign. Food trays and accumulated trash were heaped in piles on the floor in front of the closed door.

Ruth could hear a tiny voice from inside the tarped-up, trashed-out room: a barely audible "Help." She pulled back the tarp and took a peek.

Inside was a young man so skeletal he melted into the sheet below him. Ruth went inside and asked what he needed.

His momma. He was crying for his momma.

His name was Jimmy. The nurses refused to call his mom, so Ruth did, but Jimmy's mom was no help. Her son was already dead—to *her*, at least. Ruth held Jimmy's hand until he passed. Even then, the hospital staff was no help.

Ruth called funeral home after funeral home. Finally, one agreed to take Jimmy's body, paid for by an indigent fund at the hospital.

A friend gave Ruth a chipped cookie jar from his pottery shop, and Ruth put Jimmy's ashes inside. She dug a little hole herself—right next to her daddy's grave in a family plot, said a little prayer, and laid Jimmy to rest.

Over the years, Ruth cared for many others. Holding their hands, calling their families, burying their ashes in broken cookie jars in her family's plot. She arranged for housing and dug perfectly good food out of grocery store dumpsters for her guys. All of this, plus selling timeshares, often with her own little daughter in tow.

Ruth's name got around. Hospitals called her, but her church and her community were little help. Apparently, dying from AIDS didn't qualify as a good Samaritan moment. Her memoir, *All the Young Men*, needs to be on your must-read list.[1] It is not officially a book about faith, but Ruth did what she did because of Jesus.

Early in the AIDS epidemic, people were terrified—terrified of a licked postage stamp, a hug, or a public restroom. An infection with HIV (human immunodeficiency virus) was a death sentence until we had antivirals capable of bringing viral loads down to undetectable.

RETROVIRUSES

Smallpox. Black Death. The 1918 flu. Cholera. Measles. Polio. Ebola. COVID-19. Human history has been shaped by a tiny entity that isn't even a full-fledged member of the "life club."

Viruses are relatively simple things—they consist of

genetic material (either DNA or RNA) wrapped in a protein envelope. Viruses can't replicate on their own, hence their current exclusion from the "life club." A virus must hijack a host cell to do its bidding, which is "make more viruses." When a virus lands on the surface of a cell, it injects its genetic material (either DNA or RNA) into the host cell.

For our purposes, we are going to focus on RNA viruses. Common RNA viruses are those that cause influenza, COVID-19, and measles. When an RNA virus injects its genetic material into a host cell, the viral RNA stays in the host cell's cytoplasm. The viral RNA hijacks raw materials in the host cell's cytoplasm and starts building new viruses. These viruses burst from the host cell, and there you are—a body full of brand-new flu viruses, in bed with a bowl of soup and *The Price Is Right*.

There is a special category of RNA virus known as the *retroviruses*. If a virus is a retrovirus, an enzyme called *reverse transcriptase* enters the host cell along with the viral RNA.

Using raw materials in the cytoplasm of the host cell, reverse transcriptase uses the viral RNA as a template to build a double-stranded piece of DNA. This new piece of DNA, reverse-engineered from viral RNA, inserts itself into the host cell's DNA. It strong-arms its way into the host cell nucleus, settles in, and becomes one with the cell's original DNA.

And just like that—the genetic instructions to make a virus have a new, permanent home in the host cell DNA. As the cell uses its DNA to make its own products, it will also churn out viruses—viruses that can spread throughout the body, infecting more cells.

HIV is a retrovirus. When HIV infects a cell and inserts its reverse-engineered DNA into the cell's DNA, the infected cell will sometimes go into a dormant state for a long time. Current HIV treatments are highly effective at reducing the number of active viruses in the body but aren't so good at finding and destroying dormant HIV. If treatment stops, viruses reactivate—making it impossible (at this point) to "cure" an HIV infection.[2]

SCARS FROM THE PAST

June 26, 2000. "Hail to the Chief" plays as the men enter the East Room at the White House. President Bill Clinton is first to the podium, flanked by a younger and more rested Dr. Francis Collins—it's twenty years before his 100-hour pandemic workweeks. The East Room audience includes ambassadors from all over the world, and British prime minister Tony Blair joins via satellite. "Today," said the president, "we are learning the language in which God created life."[3]

The Human Genome Project, led by Dr. Collins, provided the first read-through of the entire human genome—the DNA that codes for us, the recipe to make a human. The event was momentous and deserving of its global announcement.

As scientists worldwide pored over the data, a surprising pattern emerged: 8 percent of the human genome was viral DNA.[4]

There's an interesting twist in the story of retroviruses. If a retrovirus invades a body cell (as HIV does), only the infected person carries the evidence of the infection. However, if a retrovirus infects the cells that make eggs or sperm, the viral DNA is incorporated into every egg or sperm made. If one of these eggs or sperm makes a new human, that new human will carry the viral DNA in every cell of their body. From that point on, the viral DNA is one with the human DNA, passed on to generations of offspring.

Viruses that take up residence in the DNA of another species and are passed from generation to generation are called *endogenous* retroviruses, or ERVs. Why, then, are we not constantly sick from all these viral hitchhikers in our genomes?

Human ERVs are ancient scars of viral infections in our ancestors millions of years ago. An ERV on human chromosome 17 is at least 104 million years old, maybe older.[5] Mutations, accumulated over millions of years of cell divisions, have inactivated our ERVs. They've lost their virus-y-ness and can no longer make us sick.

Only four retroviruses (including HIV) are currently known to infect humans.[6] Retroviruses are always bookended by recognizable stretches of DNA called "long terminal repeats," or LTRs. When we find LTRs in the human genome, it's a telltale sign of an ancient viral infection.[7]

ERV-W is a family of endogenous retroviruses found all over the human genome. We have identified 211 ancient ERV-W infections scattered throughout our DNA.

ERV-W is also common in the genome of chimpanzees. This is not surprising—current viruses often infect both humans and chimps. What would we expect to see if humans and chimpanzees were separately and specially created? What would we expect to see if humans and chimpanzees share common ancestry?

Here's what we see: Humans have 211 ERV-W infections. Chimpanzees have 208 ERV-W infections. There's more: Humans and chimpanzees have 205 of the infections in common. And even *more*: The 205 in-common infections are found in the same spot in both chimps and humans—same chromosomes, same locations.[8]

What are the chances? What are the chances that chimps and humans, separately, were infected by the same ERV, and these separate viral infections were inserted in the exact same spots in both the chimp and human DNA? Now do it 205 times. It's really big math, like bigger-than-all-the-universe big.[9]

What are the chances that chimps and humans, separately, were infected by the same ERV, and these separate viral infections were inserted in the exact same spots in both the chimp and human DNA?

Our genomes tell a story. Humans and chimpanzees share a common ancestor who lived around six million years ago. Before the human and chimp lineages split, the common ancestor was infected with ERV-W viruses. The ERV-W infections unique to humans and the ERV-W infections unique to chimps occurred after the split from the common ancestor.

MAKING THE HITCHHIKERS WORK FOR US

Sometimes, retroviruses do more than simply leave scars behind. Early in the story of mammals, a viral infection revolutionized mammalian reproduction. The retroviral DNA coded for a protein called syncytin, a protein used by viruses to fuse with a host cell. Mammals co-opted the cell-fusing function of the viral protein and used it in the development of a new, uniquely mammalian structure.

Built using the leftovers of a viral infection, this structure allows female mammals to carry their developing babies safely inside their bodies, rather than leaving them to fend for themselves in a nest. Syncytin proteins fuse cells from the mother with cells from the baby, allowing the exchange of nutrients, gases, and wastes: introducing the placenta.[10]

Over time, multiple versions of syncytin-coding retroviruses infected different lineages of mammals, but the end result is the same—syncytin proteins facilitate the attachment of the placenta (made by the embryo) to the uterus of the mother.

Throughout our bodies, viral hitchhikers have been put to work. ERVs play roles in the formation of stem cells, in the breakdown of carbohydrates, in immune responses, and much more.

THE MYSTERY OF THE MISSING CHROMOSOME

It's a mystery worthy of Nancy Drew. Since the days of Charles Darwin, we've been pretty sure that the great apes were the closest living relatives of humans. Comparative anatomy and fossil evidence were our best clues.

Modern genetics seemed to seal the deal—the DNA of humans and chimpanzees is more than 98 percent similar. But there was a pesky problem: Chimpanzees (as well as bonobos, gorillas, and orangutans) have forty-eight chromosomes; humans have forty-six. Quick reminder:

We have twenty-three unique chromosomes, but two copies of each.

Maybe a chromosome was lost somewhere along the way in the human lineage? Unlikely—the loss of that much important DNA would certainly be fatal.

The mapping of the human genome and the subsequent mapping of great ape genomes solved the mystery. Two chimpanzee chromosomes (12 and 13) line up almost gene for gene with human chromosome 2. Sometime after the split from the common ancestor, chromosomes 12 and 13 fused in the human lineage.

So, what's the clue, Nancy Drew? Actually, we have a plethora of clues. And better than that, we have unmistakable evidence of a fusion.

Packed into each human cell are forty-six molecules of DNA—your chromosomes. In order to grow, heal, repair, and produce offspring, cells must divide. Each time a cell divides, it makes a copy of its DNA as well, all forty-six strands of it.

Every time a cell divides, some of the DNA at each end of a chromosome erodes away. If the tips of a chromosome contained important genes, that would be very bad news. However, at each tip of a chromosome (both top and bottom) we find a long stretch of noncoding DNA. These protective caps of noncoding DNA are called *telomeres*. When cells divide, it is the telomeres that erode, not the important stuff. Eventually, the telomeres are depleted, and important DNA is damaged. With the loss of enough functional DNA, the cell dies. We experience this as the aging process.

Telomeres are easily recognizable. They are made of dense, repeated sequences, lined up in tandem.

In fact, high-density repeats are the "calling cards" of telomeres.

With our ability to map genomes, the mystery of the extra chromosome is solved. Human chromosome 2 has telomeres at each tip, just as we would expect. But right in the middle of the chromosome are two additional telomere sequences, stuck together, head-to-head, forward repeats and backward repeats.

There is one more bit of chromosome "geography" important to this discussion. Around the midpoint of a chromosome is another easily recognizable sequence called a *centromere*. Centromeres play an important function in sorting out chromosomes during cell division. On one side of the chromosome 2 fusion site is a functioning centromere. On the other side of the fusion site is a second, nonfunctioning centromere.[11]

So precise is our ability to map the genome, we can identify the exact fusion site of the two chromosomes.[12]

Humans last shared a common ancestor with chimpanzees about six million years ago. The common ancestor had forty-eight chromosomes. Chimpanzees (and all nonhuman great apes) have forty-eight. After the human lineage split from the chimp lineage, two chromosomes fused in the human lineage. We don't know exactly when the fusion occurred, but we know the common ancestor of Neanderthals, Denisovans, and *Homo sapiens* had forty-six chromosomes.[13]

ARGGHH, YE SCURVY DOGS

Prior to the eighteenth century, sailors had it rough—long voyages over stormy seas, cramped quarters, fighting rats and weevils for stale, moldy food. A few months into a voyage, sailors regularly fell sick from scurvy. They suffered bleeding, swelling, emaciation, and often, death. Scurvy killed more than two million sailors between the sixteenth and eighteenth centuries.

In 1747, a British naval surgeon carried out the first randomized clinical trial on record. Sailors sick with scurvy were divided into groups and given either vinegar, cider, seawater, sulfuric acid, or fresh citrus fruits each day. The sailors given citrus fruits made a rapid recovery.[14]

Eventually, all British ships carried a supply of lemon juice, then lime juice, for each man on board. American sailors nicknamed the Brits "Limeys."

Most mammals have a gene called GLO—a gene that allows a mammal to synthesize its own vitamin C. About forty million years ago, the GLO gene mutated in early primates, rendering it nonfunctional. At this point in primate evolution, the lemurs had already branched off on their own, and today, lemurs are the only primates with a functional GLO gene. The break in the GLO gene is in the same location in all nonlemur primates.

Early nonlemur primates were living in trees and, with their sharp color vision, were finding (and eating) colorful, vitamin-rich fruits. Broken GLO gene? No problem.

No (nonlemur) primates, including humans, can make their own vitamin C. All other mammals (except guinea pigs and a few bats) have a functioning GLO gene. Guinea pigs and some bats also have a broken GLO gene, but it is broken in a different place and in a different way than what we see in primates.[15]

OUR EGGY HISTORY

Mammals can trace their evolutionary history back to a reptile-like amniote. The earliest mammals—those closest to the base of the mammal family tree—still laid eggs. Modern monotremes (like the platypus) retain several "basal" mammal traits, like egg laying. Monotremes are a teeny-tiny exception to the mammal rule. Almost all mammals nourish their young internally, directly through the placenta.

Reptiles and their reptile cousins, the birds, have three genes for making a high-calorie, nutritious protein called vitellogenin. In egg-laying animals, vitellogenin fills the yolk sac inside the egg. Mammals, too, have the *same* three vitellogenin genes, but each one of them is broken and nonfunctional. Monotremes have one functional vitellogenin gene, but the other two are broken.

Placental mammals also make a yolk sac, but it is useless and empty. Humans have a yolk sac during the first month of pregnancy, but it never fills with yolk. By the second month, it detaches from the embryo.[16]

THE NOSE KNOWS

Humans smell really bad. Not that we are stinky, but we have a very mediocre sense of smell. Only dolphins smell worse than we do.

The largest family of mammalian genes are the olfactory receptor (OR) genes. More than one thousand OR genes code for all sorts of proteins lining the surface of the nasal cavity. We exist in a cacophony of smells—our OR genes allow us to sort it all out.

Mice are mostly active at night, so smell trumps vision. A highly charged sense of smell allows you to find your mate, your babies, your food, and warns you about hungry cats—all in darkness. Almost all the mammalian OR genes are functional in mice.

Day-active humans and other primates evolved sharp, technicolor vision. Smell took a back seat. In humans, more than 60 percent of the OR genes are nonfunctional.[17] Dolphins, who have no use for airborne odors, have lost 80 percent of their OR genes.[18]

If a gene (like those in the OR family) is important for survival, any mutation in that gene harms the animal. Only animals without the mutation survive and leave offspring. On the other hand, if a gene is not important for survival, mutations can accumulate in the gene without impacting the animal. Before long, a gene (or a whole group of related genes, like the OR group) is dead—mutated beyond functionality.

FOSSILS IN THE GENOME

Decades before we mapped the human genome, geneticist Susumu Ohno wrote, "The earth is strewn with fossil remains of extinct species; is it a wonder that our genome too is filled with the remains of extinct genes?"[19] As fossils of trilobites, dinosaurs, and mastodons tell the story of life on Earth, the human genome also holds fossils, fossils of the unfolding story of human evolution.

Pseudogenes are genes that have lost their original function, genes destroyed by millions of years of accumulated mutations. A pseudogene may be the relic of a single gene. Sometimes cells inadvertently make two copies of a gene during DNA replication, and the second, unused gene becomes a spare tire, deteriorating in the trunk from disuse. Other pseudogenes, like ERVs, are scars of an ancient viral infection. There are about twenty thousand pseudogenes in the human genome—roughly the same number of pseudogenes as functioning genes.[20]

Before modern genetics, fossils told the story of evolution. The fossil record is a nested hierarchy of morphology. The earliest animals in the record do not have heads—think about those weird Ediacarans. Eventually, we see creatures with heads, then creatures with limbs, and then, in the youngest layers of rock, we see creatures that walk upright on two legs. Evolutionary biologist J. B. S. Haldane was once asked what evidence would disprove evolution. Haldane famously snarked, "Rabbits in the Precambrian." The fossil record reveals a hierarchy, and the hierarchy is predictable.

Likewise, pseudogenes reveal nested hierarchies of relationships. Humans share more identical pseudogenes with chimpanzees than we do with gorillas, and more with gorillas than with orangutans.

Humans share almost 99 percent of our coding DNA with chimpanzees, and 96 percent when we include noncoding regions (pseudogenes and control genes). Gorillas share 98 percent with us, and orangutans share 97. Mutation rates in apes show the same nested hierarchy: humans split from chimps about 6 million years ago, from gorillas about 8.5 to 12 million years ago, and from orangutans about 9 to 13 million years ago.

We share 85 percent of our genes with mice and 60 percent with fruit flies. It's been a billion years since we last shared a common ancestor with yeast, yet we share about 25 percent of our genes. We can insert human genes into yeast cells, and they work![21]

Humans share common ancestry with all life. What's more, our DNA reveals *how* we are related and *when* we last shared a common ancestor.

BIBLICAL GENETICS

Resistance to human evolution (and evolution in general) has long focused on a missing "missing link." Show me a half-man, half-ape fossil, then we'll talk.

Evolution theory never predicts an unbroken series of fossils in *any* lineage, including humans. Fossils demonstrate the general trajectory of evolution, over time.

Fossils are gold, but DNA mapping is a game changer. It's not a real-time video recording, but the last three decades have given us the closest we've come to a smoking gun for evolution evidence. With the ability to map genomes came repeatable and consistent genetic evidence demonstrating the place of humans in the grand scheme of life.

DNA evidence is hard to ignore. DNA convicts and exonerates in a courtroom. DNA determines paternity. Order a simple swab kit in the mail, and you can identify through DNA from which part of the world your ancestors originated—and how Neanderthal you are! Genomic medicine tells us the odds of passing on a hereditary disease and can determine the best course of treatment for cancer.

The genetic science used to determine the best cancer treatment is the *same* genetic science that tells us we share common ancestry with the great apes. There's no such thing as secular genetics and biblical genetics. DNA is DNA. Yet, creationists insist that all genetic evidence be viewed with a "biblical" lens. Microbiologist and creationist Andrew Fabich calls ERVs an "evolutionary fairy tale" and rejects the implication of common ancestry out of hand: "Since we know the Word of God is true, we know that ERVs cannot be proof of evolution."[22] So, Dr. Fabich, what are ERVs? Seen through a "biblical lens," ERVs are either an artifact of a fallen world, he says, or most likely, evidence of intelligent design.

Biochemist Todd Wood is a creationist blogger and is regularly featured in *Is Genesis History?* productions. Wood likewise approaches genetic evidence with a predetermined assumption of special creation: "Given the separate origins of humans and chimps," Wood says, "we've either experienced mutations that are exactly the same as in chimps, or those 'mutations' were created that way in the beginning." Because creation is the simpler explanation, Wood concludes that the presence of human and chimp ERVs in the same locations is merely "apparent."[23]

Interesting sidenote: Wood found himself in hot water a few years ago with his fellow creationists. Wood wrote an essay in which he states, "Evolution is not a theory in crisis . . . there is evidence for evolution, gobs and gobs of it . . . it has amazing explanatory power." Wood then adds: "It is my own faith choice to reject evolution."[24]

Across the board, creationist organizations deny evolution evidence in the human genome. Inherited pseudogenes and the presence of a chromosomal fusion are simply "apparent" and are the result of purposeful design. The head-to-head telomeres of human chromosome 2 are functional genes, it's claimed, and the cryptic second centromere is nonexistent.[25]

If you're interested in a technical explanation of the genetic evidence and answers to creationist denials, "Gutsick Gibbon" is an excellent resource. Erika, the creator of the Gutsick Gibbon YouTube channel, is a paleoanthropologist and a former young-earth creationist who respectfully engages anti-evolutionists. Erika can be spicy at times, but only after spice has been served.[26]

DINOSAURS AND DNA

You don't hear it much anymore, but occasionally the existence of dinosaurs is questioned by creationists. In 2024, popular conservative political commentator Candace Owens opined about dinosaurs in an interview. Owens said dinosaurs "seem pretty fake and gay," then continued with her religious rationale for believing so.[27] Owens didn't elaborate on the sexual orientation of dinosaurs, but her views on their existence aren't unlike those of many folks in creationist quarters. Dinosaur-deniers tend to fall into one of two camps: (1) those who believe fossils are the remains of other large animals, possibly preflood, or (2) those who believe God placed the bones in the ground to test our faith—as if God pulled up layers of rock strata, threw in a few bones, and chuckled to himself, "Let them figure *that* out."

What does it say about God if our genomes look for all the world like we share common ancestry with all life? What does it say about God if our genomes were specially designed to look just like we would expect they'd look if we shared a recent common ancestor with chimpanzees? Did God chuckle to himself, throw in a few ERVs and pseudogenes, and ask us to look away? Are such deceptions consistent with the nature of God?

Why are those arguing against evidence in the human genome always religious? If the evidence against chromosome fusion or ERVs or nested relationships is so overwhelming, you'd expect at least a few nonreligious geneticists to also reject fusions, ERVs, and nested relationships.

Scientists don't talk about proof. Scientists talk about consensus. And consensus is not a vote, as if all scientists in "favor" of evolution raise their hands and the most hands win. Scientists love to prove each other wrong and will only agree with each other if there's no logical reason to disagree.

Human evolution is not true because the Science Club got together and voted it in. Rather, human evolution is true because there is consensus of the *evidence*. Common ancestry is not about one fact here or there or a popular vote, but a triangulation of evidence from multiple sources.

DISCUSSION PROMPTS

1. Humans and chimpanzees have 205 ERV infections in common. The ERV infections are found in the same location in the genomes of both species. Does this surprise you? What do ERV infections tell us about common ancestry?
2. What evidence points to a fusion of two chromosomes in the human lineage? When (approximately) did the fusion occur? What can we conclude from the fusion evidence?
3. Humans and other animals active during the day

have a keen sense of vision but a not-so-great sense of smell. Mice and night-active animals have a keen sense of smell but lack sharp vision. How does evolution explain these features?

4. Some mutations are costly, even deadly, while other mutations have no impact at all. Explain.
5. Human genes inserted into yeast cells can function. How is this possible?
6. Pseudogenes are "fossils in the genome." Explain the analogy.
7. People often trust the science when DNA identifies the best treatment for cancer or convicts (or exonerates) in court but reject the science when DNA indicates common ancestry. What accounts for such inconsistencies?
8. Creationist Todd Wood offers two possible explanations for the DNA evidence indicating common ancestry of humans and chimpanzees: Either both lineages experienced the exact mutations and viral infections, or God created humans and chimps with the appearance of common ancestry. What do you think? In your judgment, are either of these explanations reasonable? Why or why not?
9. Todd Wood wrote, "Evolution is not a theory in crisis," and there are "gobs and gobs" of evidence for evolution. Wood concluded, however, that his rejection of evolution is a personal "faith choice." What are your thoughts regarding Wood's conclusion? Is it, for you, an intellectually satisfying conclusion? Why or why not?
10. Compare the terms "proof" and "consensus" in the context of science.

9

ARE WE STILL EVOLVING?

Out of the three billion or so base pairs that make up human DNA, there is a single gene with a big job.

This gene codes for hemoglobin, the protein in red blood cells responsible for carrying oxygen to every cell in the body. Just one change, just one mutation at a single point in the gene, results in a defective version of hemoglobin.

Healthy red blood cells have a characteristic biconcave shape, perfect for squeezing into the tiniest of blood vessels to deliver the oxygen payload carried by hemoglobin. Defective hemoglobin changes the shape of a red blood cell. Instead of smooth and biconcave, the cells are hard and sticky and are C-shaped, like a crescent or sickle.

We have two copies of every gene—one from mom, one from dad. If one copy of the hemoglobin gene is operational, enough functioning red blood cells are made to keep a person healthy. A carrier of one mutated hemoglobin gene may never know it.

However, if a person has two copies of the mutated gene, there's big trouble. Red blood cells get sticky and clump together and block vessels. People with sickle-cell disease have excruciating pain and often suffer strokes. Sickled cells are short-lived, so sufferers are anemic, exhausted, and prone to serious infections and kidney failure. Before modern medicine, children with sickle-cell disease often died before their fifth birthday.

British physician Anthony Allison grew up in the highlands of Kenya, overlooking the Great Rift Valley. Allison returned to Africa in the 1950s with the intent of studying the role of ABO blood groups and sickle-cell disease in East African people.

Allison collected blood samples from almost five thousand people. In some areas, 40 percent of the population were carriers of a single mutated hemoglobin gene. Before long, Allison noticed a curious phenomenon—most carriers were clustered in warm coastal areas. In Kenya, tribes living on the coast or near Lake Victoria had carrier rates of 20 percent, while those in the Kenyan highlands where Allison grew up had carrier frequency rates of less than 1 percent.[1]

The warm, wet areas of Kenya are breeding grounds for the *Anopheles* mosquito, carrier of the deadly parasite that causes malaria. Is there a connection between carriers of the mutated hemoglobin gene and malaria? Allison wanted to know.

Allison soon had his answer: Carriers of the mutated hemoglobin gene are resistant to malaria. Carriers produce just enough defective hemoglobin to make their blood cells inhospitable to the malaria parasite, but they produce enough functioning hemoglobin to stay healthy. Carriers of one mutated gene are therefore resistant to malaria *and* do not have sickle-cell disease. People with two copies of the mutated gene are also resistant to malaria, but they have a bigger problem: a debilitating, potentially lethal disease.

Anthony Allison discovered the first undisputed example of natural selection in humans.[2]

A FINISHED PRODUCT?

The story of the mutated hemoglobin gene seems to flout the rules of evolution—why would such a potentially lethal gene be so prevalent in some populations? Lethal genes should become rarer, not more common.

In areas where mosquito-borne malaria is a problem, resistance is gold. In a single year, more than 263 million people will have malaria, mostly young children and pregnant women. Of those contracting malaria, 597,000 will die.[3]

Historically, those suffering from malaria either left far fewer offspring than others or died before having children. People with one copy of the mutated hemoglobin gene are generally much healthier—no sickle-cell disease and a strong resistance to malaria. People with one copy of the mutated gene live longer and have more children. In other words, there is a strong selective advantage to having one copy of the lethal gene, so it spreads in the population.

With so many carriers in a limited geographic area, it won't be long before two carriers have children together. There is a 25 percent probability that two carriers will have a child with sickle-cell disease, but that is a probability—not a guaranteed outcome. Five children produced by two carriers might all have sickle-cell disease, or none may have it, or somewhere in between—the probability is 25 percent for each pregnancy.

Human evolution is an often-ignored single section in a textbook or a quick reference in a broader discussion. It's easy to conclude that we're a finished product and the story is over. But humans never stopped evolving, and we are evolving still.

The hemoglobin mutation arose in Africa in a single person around 7,300 years ago.[4] Without a strong selective advantage, the mutation meandered through the population, never really gaining much steam.

Environmental changes drive natural selection. When change comes, a trait boosting fitness in the new environment will take off in a population. A few thousand years after the hemoglobin mutation arose, change came to eastern and southern Africa. The Bantu people, originally in central and western Africa, expanded to the east and south, bringing a new way of life. The Bantu were farmers, and as they spread, they cleared woodlands for farmlands on a massive scale.[5]

As the woodlands cleared, mosquitoes moved in. As farming (and farmers) increased, human hosts for the malaria parasite increased.

Wherever malaria mosquitoes were problematic, the hemoglobin mutation increased. In southern Africa where there are fewer malaria-carrying mosquitoes, the mutation is rare. Eventually, migrants with the mutation moved into the Near East, into Europe, and into India. In areas of the world where malaria is a problem, we find carriers of the mutation.[6]

GOT MILK?

Supermodel Naomi Campbell. David Beckham. Serena Williams. Taylor Swift. Bart Simpson. The "Got Milk?"

milk-mustache campaign made it cool to drink milk and appear in print with the evidence.

A lot of effort went into an ad campaign for a product that makes most people in the world have diarrhea.

Mammal babies (including humans) can drink and digest milk. In fact, producing milk for babies is a distinguishing trait unique to mammals. But after infancy, milk can make mammals sick. Despite popular assumptions, it's not a good idea to give a saucer of milk to an adult cat.

Milk contains a sugar called lactose. Lactose molecules are too big to be easily digested, so an enzyme called *lactase* breaks lactose into smaller, more digestible sugars. Most humans, however, stop producing lactase sometime between childhood and early adult years. Lactose passes undigested into the large intestine where lactose-loving bacteria live. The bacteria break down the lactose, and in the process produce painful gas, cramps, and diarrhea.

However, 35 percent of humans continue to produce lactase throughout their lives and have no problem consuming milk or dairy products. And interestingly, lactase persistence is concentrated in limited areas around the globe—areas known as "lactase hot spots."

The capacity to consume milk, ice cream, and milkshakes without getting sick is a story of (relatively) recent human evolution.

Around ten thousand to thirteen thousand years ago, a mutation occurred in the lactase "on-off" switch. The mutation kept the switch stuck in the "on" position, and carriers of the mutation continued to produce lactase throughout their lives. Carriers, however, never noticed. Cultures at the time were almost exclusively hunter-gatherers, so dairy was not a part of the diet.

As agriculture spread into Europe from the Fertile Crescent, farmers domesticated animals as well. By seven thousand years ago, we find archaeological evidence of dairy herds and milk fats in cooking vessels.[7] During the long, cold, dark winters of northern Europe or in times of crop failure and famine, milk was a protein-rich, fat-rich food source. The lactase-persistence mutation provided a selective advantage: People with the mutation consumed calorie-rich dairy foods without a problem and were therefore healthier and had more children. People without the mutation got sick and dehydrated when they were already on the brink of starvation.[8]

Early European farmers were not the only ones who valued dairy animals. African farmers were herding dairy animals and consuming milk around the same time. Three different lactase-switch mutations have been identified in African populations.[9] Being lactase-persistent was such a selective advantage in dairying cultures, mutations in the on-off switch occurred at least four different times. In Europe and in Africa, human evolution followed a parallel track, driven by natural selection.

In parts of the world where dairying was not historically practiced, there was no selective advantage in carrying the mutation. Before long, the mutated gene disappeared or fell silent in these populations. In modern human populations, lactase persistence is dependent on early agrarian ancestors.

THE VIEW FROM THE TOP

The view from the "roof of the world" must be spectacular—I can't imagine I'll ever see it firsthand. The first confirmed climbers to summit Mount Everest did so in 1953, followed by many others—some successful, others not so much.

Mount Everest is just a kid, geologically speaking. Everest was formed fifty to sixty million years ago when India crashed into Asia, shoving up the highest mountains (from sea level) on earth. At 29,032 feet, the summit offers one-third the air pressure that exists at sea level. At this pressure, humans gasp for oxygen. Climbers are at risk for swelling in the brain and lungs and for blood clots. At this altitude, pregnant women are at risk for miscarriage and pre-eclampsia.

But not all humans are at risk. The first permanent settlements on the Tibetan plateau emerged between five thousand and nine thousand years ago, but humans have lived in the area for at least thirty thousand years.[10] Modern Tibetan highlanders thrive in the low-oxygen environment.

Researchers analyzed the DNA of Tibetan highlanders for clues. Specifically, they were looking at a gene that regulates the body's use of oxygen. There is a version of this gene found only in Tibetan highlanders and, in very low frequencies, among Han Chinese. This version of the gene makes Tibetan highlanders very efficient at using the limited oxygen of extreme elevations. Among other things, Tibetans take more breaths per minute than people at sea level.[11] Although this gene has not been found in any other living group of humans, it is present in one other source: the DNA of Denisovans, an extinct human species closely related to the Neanderthals.[12]

Tibetans aren't the only people living at the top of the world. People living on the Andean plateau have adapted to low-oxygen air by increasing the amount of hemoglobin in their blood. People living in the Ethiopian highlands have also adapted to low-oxygen air. Ethiopian highlanders don't breathe more rapidly like the Tibetans, nor do they make extra hemoglobin like the Andeans. The Ethiopian highlanders are not as isolated as the Tibetans and have genes from outside Africa, making it more difficult to identify the source of their genetic adaptation.[13]

The first humans settling in extremely high elevations faced survival challenges. Those with any type of gene variant making it easier to live in low oxygen would be healthier, more fit for the environment, and more likely to leave offspring. Over generations, beneficial adaptations increased in the population until all in the population had the trait or traits. The population evolved.

Notice, all three populations—the Tibetans, the Andeans, and the Ethiopian highlanders—are adapted to low oxygen, but all three solved the low-oxygen problem in different ways. This is called *convergent evolution*, and we see it throughout the tree of life. Dolphins (mammal), sharks (fish), extinct ichthyosaurs (reptile), and penguins (bird) are very different animals, but they all live in an ocean environment. All four solved the problem of living in water in a similar way: a streamlined body with flippers or fins. All four arrived at the solution via different pathways yet solved the problem in a similar way.

Y OH Y

There is a gene for red-green color vision found on the X chromosome. Two versions of this gene exist in the human population: one functioning version, and one rare, nonfunctioning version. If a female inherits the nonfunctioning version on one of her X chromosomes, lucky her—she has a second X chromosome. One functioning color vision gene is enough to have normal color vision. Males aren't so fortunate. If a male happens to inherit the nonfunctioning version on their one and only X chromosome, they have color vision deficiency. Men with color vision deficiency far outnumber women with color vision deficiency.[14]

All chromosomes are paired. In each pair, one is from mom, one is from dad. For twenty-two of those pairs, there is correspondence of genes, all down the line. For each gene on the chromosome from mom, there is the same gene (or a version of it) on the chromosome from dad.

Not so with chromosome pair 23, the X and Y. The X chromosome is much larger than the Y chromosome. It's not surprising, therefore, that the X chromosome carries many genes for which there is no corresponding gene on the Y. X has about nine hundred genes, Y has a paltry fifty-five. Sex-linked or X-linked traits—traits that are far more common in males than females—include color vision deficiency and certain types of hemophilia.

The Y chromosome has the fewest genes of any chromosome. What's more, many Y genes are duplicates and most are inactive. One gene on the Y gets all the attention: the SRY gene (**S**ex determining **R**egion on the **Y**). At about twelve weeks of gestation, the SRY kickstarts the development of testes, which in turn direct the development of typical male genitalia.

If the SRY gene is missing or inactivated for some reason, the baby will develop as a female with external female genitalia, a vagina, uterus, and Fallopian tubes, but will be genetically XY—a condition known as Swyer syndrome.[15] Disorders/differences of sexual development (DSD), like Swyer syndrome, are rare but real. Controversy arose in the 2024 Olympics when a female participant in women's boxing likely had a DSD.

Despite the important SRY gene, the human Y chromosome shows signs of being a degraded chromosome near the end of its life.[16] In the last 180 million years, the Y chromosome inherited by humans has lost 97 percent of its genes. At this rate, the Y chromosome could disappear in as little as 5 million years.[17]

The degradation of the Y chromosome is a fascinating glimpse into human evolution.

Part of the problem is the size disparity between the X and the Y. During the special kind of cell division (called *meiosis*) that creates eggs and sperm, paired chromosomes come together and swap out bits of DNA. Chromosome pairs cross over each other and make trade-sies. The Y, however, is so much smaller than the X, crossing over is rare. It's hard for the Y chromosome to get rid of mutations or deletions or any other sort of DNA junk. Compounding the problem is the fact that, unlike egg production, sperm production is always full speed ahead. The cells that make sperm are always dividing, increasing the chances of mistakes and mutations.

Is this bad news for the guys? Will human males be extinct in the future? Probably not. We know of at least two rodent groups who have lost their Y chromosomes. Males still exist in these species, so apparently another gene has (or other genes have) evolved to perform the function of the SRY.[18]

The degradation of the Y chromosome is a fascinating

glimpse into human evolution. We have clues about what the Y was like 100 million years or so ago, but we don't know what its evolutionary fate will eventually be. What we have is a look at evolution in action, in the middle of the action.

DRIVERS OF EVOLUTION

Sometimes factors driving human evolution are quite accidental. In the seventeenth century, a small group of Dutch colonists settled on the Cape of Good Hope. The settlers just happened to carry the gene that causes Huntington's disease with unusually high frequency—a frequency far greater than in the Dutch population as a whole. This small settler group were the founders of the Afrikaner population in South Africa. The Huntington's gene is found in much higher frequencies in modern Afrikaners than in other populations, an evolutionary mechanism called "founder effect."[19]

Sometimes behaviors drive evolution. A large-scale genetic study in the United Kingdom and Northern California looked at a nicotine receptor gene that makes smoking very addictive. The gene was prevalent in the children of British men who died young, and so presumably, it was also prevalent in their deceased parents. Many of the men who died young reached adulthood in the 1950s, a time when many British men smoked heavily. In Northern California, however, the gene's frequency did not vary across age groups, presumably because few (regardless of age) smoked heavily, and the gene did not impact their survival. "As smoking habits have changed, the pressure to weed out the allele has ceased," said Joseph Pickrell, director of the study.[20]

As we saw with malaria resistance and altitude adaptation, environmental pressures are the strongest drivers of human evolution. Under all their dark fur, chimpanzees have pale skin, so most likely, the common ancestor of humans and chimps also had pale skin.

As the human branch diverged from the common ancestor, body hair diminished, exposing skin directly to the sun. Dark skin protects cells from damage by the sun's radiation. Ultraviolet radiation also breaks down folate in blood vessels in the skin. Folate deficiency in women is associated with a greater risk of birth defects. Humans arose in sunny Africa, and as we evolved, we increased the melanin in our skin. Where sunlight is almost constant, there is strong negative selection against pale skin. By 1.2 million years ago, all humans had dark skin.[21]

Some ultraviolet light is needed to produce vitamin D. As humans moved around the globe, they often settled in northern latitudes where sunlight is limited. In northern lands, the environmental pressure for dark skin diminished and lighter shades evolved. Interestingly, indigenous populations in northern latitudes with diets rich in vitamin D did not evolve lighter skin.

BAD DESIGN AND ENGINEERING FAILS

Mosquitoes kill, but one single change in a gene bestows malaria resistance. A good thing, right? A dandy design! With a tweak at just the right spot in the DNA, a loving intelligent designer heroically saves lives.

Not so fast. It's not a surprise that a gene conferring malaria resistance proliferates in areas where malaria is a problem. But with high frequencies in a limited geographic area, it won't be long before two carriers have children. Children who inherit two copies of the mutated gene are resistant to malaria, but at a terrific cost: sickle-cell disease. People with sickle-cell disease suffer, and without medical care, often die in childhood.

Crediting the malaria-resistance mutation to a loving choice by an intelligent designer demands consideration of the other side of the coin: the 300,000 children born each year with sickle-cell disease.

Intelligent design is the more science-y version of creationism. Intelligent design isn't an idea separate from creationism; rather it is a way of explaining it. Intelligent design says that life was micromanaged at each step by a designer. The designer is seldom identified as the God of the Bible, but without a doubt, that's the implication.

Advocates of intelligent design creationism see perfection in the "design" of the human body. Design apologetics is often accompanied by awe-inducing descriptions of structures or processes that simply "aren't possible" from a blind process like evolution. The *only* explanation is special intervention and micromanaging by God.

In reality, the human body testifies to a history of jerry-rigging, improvising, and less-than-ideal setups.

Our Back

Anthropologist Bruce Latimer calls the human back an "engineering nightmare": "If an engineer were given the task to design the human body, they would never have done it the way humans have evolved," says Latimer.[22]

Humans inherited their backbone from animals who walked on all fours. On all fours, the body's weight hangs from a horizontal spine. Humans took that spine and stood it upright, throwing everything out of balance. As our ancestors shifted to upright, bipedal walking, we evolved an S-shaped curve in the spine. Our S-shaped spine is flexible and allows us to stand up, walk, and run—but at a cost. Patching up a horizontal spine to make it work for an upright bipedal human means wedge fractures, scoliosis, damaged discs, and debilitating nerve pain.[23]

Our Throat

Eating and breathing come so naturally, so autonomously, it's hard to imagine them as deadly. Individually, not so much. But the windpipe (the trachea) and food pipe (the esophagus) share a common passageway. One false move, and you'd better hope your buddy knows the Heimlich maneuver.

The epiglottis is a small flap that, if all goes well, seals off your trachea when you swallow food. If the epiglottis is late on the draw, food enters the windpipe, and you choke. Choking, of course, can be fatal, but this flawed design has other health consequences. Gastric reflux sends acid up and out of the stomach that can enter the lungs through the common passageway and damage delicate lung tissues.

There's no need for the windpipe and the food tube to share common space, says Luke Janssen, physiologist and professor of medicine at McMaster University. If humans are the result of a micromanaged design, why not side by side tubes? We might have a slightly thicker neck, but we wouldn't know any differently.[24]

Childbirth

For most of human existence, childbirth was a deadly proposition for both mother and baby. Humans evolved from animals who laid eggs or gave live birth through the pelvis, and we inherited that setup.

Humans, however, evolved a big brain housed in an oversized head. Our big-headed babies are birthed through the pelvis, which is limited in width to allow bipedal walking. Often babies get stuck, and in the past, both mother and child suffered horribly and died. A designer could have rerouted birth through the lower abdomen, bypassing the narrow pelvis. Human surgeons designed a similar solution to bypass this flaw—the C-section.[25]

A WORK IN PROGRESS

Our bodies testify to a long history of structures and processes inherited from earlier ancestors. Evolution is not goal oriented, and evolution is definitely not perfection. If a trait is "good enough" for survival and results in offspring in the next generation, the trait sticks around. As we adapt to changing environmental challenges, we are evolving still. Evolution has no end goal in mind.

The quality most often associated with evolution is "survival of the fittest," but Charles Darwin did not coin the phrase. Darwin always preferred his original descriptor—natural selection. Evolution is, in fact, *not* survival of the fittest. A creature does not have to be the strongest, fastest, smartest, or the very best at anything in order to survive. Evolution is not survival of the fittest; evolution is survival of the fit *enough*. If you can survive long enough to find a mate and leave offspring, congratulations! You're fit.

Imperfection is the trademark of evolution. Imperfection is what we'd expect if humans evolved from earlier forms over millions and billions of years. If humans evolved, we'd expect to see processes and structures that are modified versions of older ones—and that is exactly what we see.

Evolution may be innovative, but it won't always be elegant. Evolution is problem solving. Evolution results in *a* way of solving a problem, but it might not be the *only* way or even the best way. After millions of years of common ancestry, we expect to find inelegant solutions to problems throughout the body.

Imperfections, jerry-rigging, and workarounds are not what we'd expect from a micromanaging designer, and that's a problem for advocates of intelligent design creationism like Michael Behe. At the end of the day, bad design is shrugged off as an unknowable mystery: "Features that strike us as odd in a design might have been placed there by the designer for a reason—for artistic reasons, for variety, for an as of yet undetectable purpose, or some unguessable reason—or they might not. Odd they may be, but they may still be designed by an intelligence."[26]

For other design proponents, it's not a mystery, it's a simple one-or-the-other situation: humans are either a "cosmic accident" or the result of an intentional act by a "master designer-engineer."[27]

Around three thousand years ago, an Iron Age poet exclaimed, "I praise you because I am fearfully and wonderfully made!"[28] The poet boasted in his creation, yet he knew nothing of the clotting cascade or the Krebs cycle or the unnecessarily circuitous and flawed route taken by the laryngeal nerve, thanks to our fishy beginnings. Does accepting our less-than-perfect bodies require us to jettison all notions of fearfully and wonderfully and specially knit together? Should we be sad?

We are wonderfully made, but not because we are an engineering perfection. It's about our function, not our formation. Despite all the bad designs and workarounds, we live.

DISCUSSION PROMPTS

1. A lethal gene should become rarer, not more common. However, the gene for sickle-cell disease continues to survive in certain populations. Explain.
2. "Lactase hot spots" are areas in the world where most of the adult population can drink milk without getting sick. What historical/cultural factor influences lactase hot spots? Why was lactase persistence an evolutionary advantage in these areas?
3. Regulating the use of oxygen at high altitudes and lactase persistence have evolved multiple times and in multiple locations around the globe, via different genetic mutations. What explains this phenomenon?
4. Human chromosome pair 23 is unlike the other twenty-two chromosome pairs. How are the X and Y different from the other chromosome pairs?
5. In your opinion, is the malaria-resistance gene a purposeful design? Why or why not?

6. How have environmental pressures driven the evolution of human skin color?
7. Anthropologist Bruce Latimer calls the human back an "engineering nightmare." Does this description offend you? Why or why not?
8. Evolution is not survival of the "fittest." Evolution is survival of the "fit enough." Explain.
9. Aspects of the human body that are odd or are not efficient are often explained as "artistic choices" by an intelligent designer. Sometimes, "bad design" is explained as a choice by an intelligent designer for an undetectable or unknowable reason. What are your thoughts?
10. "Fearfully and wonderfully made" is a cherished description of humanity by many people of faith. Before you read this book, what did this description mean to you? Do evolutionary origins change your understanding of this phrase?

10

GRANDPA ADAM AND GRANDMA EVE

Around 180,000 years ago, a woman lived in a green and flourishing wetland in what is now sub-Saharan Africa. She was a mom, and we know she gave birth to multiple children. We don't know if she had any sons, but we know she had at least two daughters.

Meet Eve—the mother of us all.

She's not, however, *that* Eve. Our African Eve was not the first woman, nor was she the only woman living at her time. She is Mitochondrial Eve, the most recent common female ancestor of all humans alive today.

Mitochondria are the structures in our cells that use oxygen to extract energy from our food. Mitochondria were once free-living bacteria until they took up residence in a bigger cell a few billion years ago. Evidence of their free-living history is the mitochondrion's tiny little chromosome—its very own DNA.

All our mitochondrial DNA is inherited from our biological mother. Sperm have mitochondria too, but they're all located in the tail of the sperm, powering it along on its precarious journey to fertilization. Only the sperm nucleus with its motherlode of DNA enters the egg. All mitochondria in the newly fertilized egg are those already present in the egg cell of the mother.

Now you know: You have a tiny bit more DNA from your biological mother than from your biological father. You aren't fifty-fifty.

We can map mitochondrial DNA, and we know how fast mitochondrial DNA mutates. Using these metrics, we can place Mitochondrial Eve about 180,000 years ago, well within the timeline for the emergence of *Homo sapiens*.

Here's a simplified way of understanding Mitochondrial Eve: If all people on Earth disappeared except for you and a cousin who is the child of your mom's sibling, your maternal grandmother would be the most recent common female ancestor of all humans, at that time. Your grandmother would be Mitochondrial Eve.

Other women were alive at the time of Mitochondrial Eve, but none of these women had descendants who made it all the way to today. And the Mitochondrial Eve "crown" can change heads. If the female lineage of Mitochondrial Eve dies out, the crown is passed to another. As populations change over time, Mitochondrial Eve changes.

Likewise, the Y chromosome is only passed from fathers to sons. "Y-chromosome Adam," the most recent common male ancestor of all modern men, lived in Africa around 210,000 years ago.[1] And like Mitochondrial Eve, Y-chromosome Adam changes as populations change.

So, it's true—we all descend from one woman and all men descend from one man. Problem is, they lived thousands of years apart and likely not in the same place—and you thought it was hard to connect on Bumble.

Mitochondrial Eve and Y-chromosome Adam are only responsible for a tiny part of our genomes. The X and the Y chromosomes are just two chromosomes out of our complete catalogue of forty-six. Our species, *Homo sapiens*, arose gradually from earlier hominids. Humans evolved as a population, with multitudes of individuals contributing to the modern human genome. It would be impossible to identify *the* first man and *the* first woman.

OUT OF A POPULATION

The amount of genetic variation between two modern humans is only about one-tenth of 1 percent. In other words, we are 99.9 percent genetically identical to each other. That's not terribly surprising: *Homo sapiens* have only been around as a species for 300,000 years. We're relative newcomers in the big scheme of things.

We know the mutation rate in human DNA (about one mutation every seventy-five million bases per generation), and we know by what percentage we differ from each other (0.1 percent). Modern humans must have emerged from a population no smaller than fifteen to twenty thousand individuals.[2]

Immediately, you see the problem. Twenty thousand is not two. A literal Adam and Eve as the actual parents of all humanity requires an alternate explanation.

Intuitively, we know that small, isolated populations have very little diversity. If, for example, a small village was isolated from other populations over several generations, everyone in the village would be some degree of cousin. There would be little genetic diversity in the village; small populations mean low diversity.

But if we look at a very large population—say the population of the planet—we see many variations and lots of diversity. Large populations mean high diversity. DNA evidence supports human origins in a population of thousands, not two.[3]

Maybe there's an alternative explanation. Maybe human DNA is not as it appears to be. Maybe, thousands-of-years, mutation-driven diversity is only an illusion. Maybe, as creationist Brian Thomas says, God prepackaged encyclopedias of coded instructions into Adam and Eve "right from the start."[4]

Did God front-load Adam with highly variant DNA? Was Eve, being made from Adam's rib, front-loaded with DNA identical to Adam (excepting the Y, of course)? Are these reasonable assumptions? Computational biologist Stephen Schaffner wanted to know.[5]

Any DNA variants present in a founding pair of humans—parents of us all—should be widely shared among individuals around the globe, says Schaffner. These variants would be "high-frequency" variants because they are found in many genomes and are spread throughout the world.

On the other hand, rare, low-frequency variants—those found in only a few genomes—would be the result of mutations occurring in the generations following Adam and Eve.

The 2003 Human Genome Project was the biological equivalent of the moon landing. Initially, 90 percent of the human genome was mapped. In subsequent years, not only have we sequenced 100 percent of the genome, but we now have hundreds of thousands of complete human genomes available for study.

From this wealth of information, what do the data show? What does the genetic evidence indicate? Does the human genome indicate front-loading of already-diverse DNA?

Not at all. The evidence indicates that *all* of our genetic variance comes from mutations. All evidence points to humans evolving in a population of thousands, going back at least a million years.[6]

Population genetics, the branch of science that studies the genetic composition of populations and how they change, isn't unique to the study of human evolution. The same science that tells us humans evolved from a population of thousands informs the management of wildlife and marine life. Population genetics is critical in the science of forensics. In medicine, population genetics informs our understanding of disease and response to therapies.

All evidence points to humans evolving in a population of thousands, going back at least a million years.

The science that says humans evolved within a population of thousands about a million years ago is the same science used to study everything from epidemics to ecology. We can't dismiss population genetics when applied to human evolution and happily accept its use in medicine, wildlife management, and environmental genetics.

TRYING TO MAKE IT ALL FIT

Twenty thousand is not two. The genetic diversity we see in modern humans either is as it appears to be—the result of at least a million years of evolution—or, by some sort of supernatural intervention, it is *not* as it appears to be.

The motivation for an alternative explanation for the genetic diversity we see in modern humans is always theological. The goal of a nonevolutionary explanation is either (1) to preserve a literal reading of the creation stories or (2) to support a theological position regarding original sin and the "fall" or, usually, (3) both.

No one wants to be perceived as antiscience. In order to conserve a literal Adam and Eve story or to support a particular theology of original sin or the "fall," or both, advocates offer different types and degrees of divine intervention in the human genome.

FRONT-LOADING

Young-earth creationists approach the study of human genetics with a nonnegotiable foregone conclusion: The earth (and thus humanity) is only six-to-ten-thousand years old.[7] This poses a special challenge—modern levels of genetic diversity must accommodate a time frame far shorter than the evidence indicates. Preeminent young-earth creationist organizations all agree: The solution is created diversity. By divine intervention, an encyclopedic collection of genetic information was front-loaded into Adam and Eve on the sixth day of creation.[8]

Some creationists take the front-loading solution and dial up the volume—way up. John Sanford and colleagues not only support created diversity in the nuclear DNA of the first pair, but in addition, they propose that *each* gamete-producing germ cell in Adam and Eve would have its own customized DNA. As a result, each sperm and each egg produced by Adam and Eve would be genetically unique.[9]

A historical reading of Genesis also means making Eve with a rib taken from Adam. With the exception (somehow) of her second X chromosome, a woman made from tissue literally removed from Adam would be a clone of Adam. Eve as an Adam-clone severely limits diversity in the primordial pair. Once again, divine intervention solves the problem: God cloned Adam's DNA, then destroyed Adam's Y, then doubled Adam's X, and just like that—we have Eve.[10]

Fazale Rana is a creationist, but the "old-earth" sort. In an old-earth model, the days of creation represent long periods of time, often termed "day-age" creationism. Within an old-earth time frame, Rana still affirms Adam and Eve as the first two humans, instantaneously and specially created from scratch, with front-loaded genetic diversity.[11]

THE BOTTLENECK PROBLEM

Created genetic diversity in two (and only two) individuals poses a fundamental problem as the population grows. When populations undergo an extreme reduction in size, the result is often extinction.

Soaring above the mountains in central Europe is a massive bird with a nine-foot wingspan, a colorful history, and an undeserved bad reputation. Bearded vultures are scavengers who primarily feed on bones—enormous bones, often swallowed whole. It was long (and wrongly) believed that these birds would soar overhead, swoop down, and carry off a lamb or even a small child. Fear drove hunting to extinction, and the last wild bearded vulture was killed in 1913.

A small population of bearded vultures live in captivity and are slowly being released into the wild. By 2003, about 120 birds lived in zoos and breeding centers. Still, conservation biologists are concerned that there is not enough genetic variability to sustain the population over the long term.[12] Lack of genetic diversity leaves a population without adaptations needed for survival in a changing environment. In small populations, harmful mutations are concentrated and beneficial traits don't spread.[13]

In the nineteenth century, one million people in Ireland died during a devastating potato famine. Over a seven-year period, most of the potato crop (the primary food source for poor farmers) was lost to disease. Irish farmers planted "lumper" potatoes—a cloned variety with little to no genetic diversity. In 1845, *Phytophthora infestans*, a fungal infection, spread rapidly throughout Ireland. Lumpers had no genetic resistance to the disease, and, because they were clones, there was widespread crop failure.

Lack of genetic diversity often leads to extinction. A population emerging from an extreme bottleneck of two humans would not likely survive.[14]

To his credit, Fazale Rana acknowledges conservation biology studies and the survival problems created by launching humanity with a population of only two. Rana's solution? Adam and Eve's DNA was "pristine" and had no harmful mutations.[15]

To the extreme-bottleneck-within-an-old-earth framework, Ann Gauger and Ola Hössjer add a bit of drama. Gauger is a senior fellow at the Discovery Institute, the flagship organization of the "intelligent design" precinct of creationism, and Hössjer is a Swedish mathematician. The two collaborated on a mathematical model for a founding pair.

Imagine, say Gauger and Hössjer, in a world of hominins, a pair are inexplicably separated from the others in an isolated gorge or valley. Or, even *more* inexplicably, all hominins on Earth except two are wiped out.[16] Lots of big math and lots of graphs later, Gauger and Hössjer conclude: Well, you can't prove it didn't happen. Inexplicable natural disaster notwithstanding, the problem of species survival following an extreme bottleneck remains.

MATH PROBLEMS

The mathematical methods used to determine the genetic diversity and the rate of mutation in a modern population are well established.[17] If scientists use these methods, ancestral populations can be estimated. Most creationist organizations acknowledge these methods as widely accepted—but only when used in modern populations.

Population genetics must therefore be attacked on other fronts. Fazale Rana thinks the math is unreliable at best but is probably wrong. Because it is impossible to *precisely* count early populations of humans like we can count modern populations of sheep or whales, Rana considers the math of population genetics "unvalidated."[18]

Here's another math attack: The rules have changed. We cannot assume that mutation rates and generation times in the past are what we observe today. And if our assumptions are wrong, say creationists, we can't trust the math.[19] "Molecular clock dating," says Elizabeth Mitchell, "*sounds like* something very precise and reliable—is in fact built on a series of unverifiable assumptions and circular reasoning. . . . And if the assumptions on which the molecular clock is based are not reliable, then all times derived from the clock are likewise invalid."[20]

CONCORDISM

Many Christian apologists look for ways to merge their theology with modern science, a practice known as *concordism*. Typically, concordists (1) pry science meaning from a biblical text, or (2) interpret Scripture in light of modern science, or (3) some combination of both.[21] Interpreting Genesis 1:1 as "the big bang" and the "day-age" view of the creation week are textbook examples of concordism.

Concordism is often infused with technical jargon and scientific terminology. Two prominent concordists, Joshua Swamidass and William Lane Craig, tackled the question of human genetic diversity.

PROBLEMS OF DEFINITION

Can a theology dependent on a recently created, made from scratch, father-of-us-all Adam coexist with empirical genetic evidence?

Computational biologist Joshua Swamidass gives it a solid try. In his book *The Genealogical Adam and Eve*, Swamidass hypothesizes a recent creation (ten thousand years ago or less) of Adam (from dust) and Eve (from a rib) consistent with the observed genetic diversity in modern humans.[22] What's more, Swamidass claims that his recent and specially created Adam and Eve are the ancestors of everyone.

It's all in how you define "ancestor," says Swamidass.

Through providential guidance, says Swamidass, humans evolved, sharing ancestry with all primates. In a unique act of creation less than ten thousand years ago, God stepped in, made Adam and Eve, and settled them into a paradise garden, exactly as described in Genesis. Then: the tree, the snake, the fruit . . . and out you go.

Adam and Eve are subsequently dropped full-grown into the big gene pool of humanity that existed outside the garden. Adam and Eve and all their descendants swam around in the big pool, and voilà!—genetic diversity.

A bit of creative terminology is required at this point. Swamidass redefines Adam and Eve not as the progenitors of all humanity but instead as our "genealogical relatives." Adam and Eve are *not* our great-great (many times) grandparents but are instead ancient residents of our family tree, like great-great (many times) aunts and uncles.

Swamidass's explanation for genetic diversity in modern human DNA is a hybrid story with the best of all worlds: We get the Sunday school version of Adam and Eve, and we also get to believe modern genetics. Swamidass calls his scenario the "genealogical hypothesis."[23]

A hypothesis, however, is not simply a "guess." It's not an "educated" guess, either. In science, a hypothesis is a possible explanation for something observed. The strength of a hypothesis is in its ability to be tested.

I might, for example, observe that people often take vitamin C to "prevent" the common cold. A possible explanation for this observation is this: Vitamin C actually prevents the common cold.

Now this is where science kicks in: A valid hypothesis must be *testable.* We must be able to design an experiment to see if our hypothesis can be supported—or not. We can give various dosages of vitamin C to some people and no vitamin C to others. We set a standard time frame. Finally, we compare the groups, and we have an answer (at least for the people in our study).

In other words, a valid hypothesis (as scientists use the term) must be *falsifiable*. If there is no way to test or disprove a claim, the claim is not falsifiable.

Swamidass's "genealogical hypothesis" is not falsifiable, so it is not a hypothesis in a science sense. A special, *de novo* creation of two people is not a claim that can be tested. Swamidass therefore focuses on the plausibility of two people being somewhere in everyone's family tree. His goal is to keep his theology intact while maintaining his credibility as a scientist. However, the "genealogical hypothesis" is unfalsifiable and as such is legitimate as a *belief* but not as science.

But what about the Bible part of the "genealogical hypothesis"?

No doubt, Swamidass is a concordist: He wants a hybrid best-of-both-worlds Adam and Eve scenario. Although he wants the Sunday school version *and* modern genetics, the "genealogical hypothesis" is not internally consistent with the Genesis account. Adam and Eve as "family tree" but *not* the parents of us all tells a completely different story.

THE TIMELINE

Christian philosopher and apologist William Lane Craig (*In Quest of the Historical Adam*) picks up the concordist challenge and kicks the can down the road—in the opposite direction.[24] Craig takes us back to pre-*Homo sapiens* hominins, traveling far past the six-to-ten-thousand-year mark.

Using Gauger and Hössjer's calculations, Craig places Adam and Eve 500,000 years in the past (352–53). Through divine intervention, Craig presumes a biological and spiritual remodeling of a pre-*sapiens* male and a pre-*sapiens* female. Thus, the newly remodeled hominins are promoted to the status of *human*. Such a promotion requires genetic modifications, says Craig—most likely in the eggs and sperm that made Adam and Eve. It is important to Craig that Adam and Eve be human "from the moment of conception" (376–77).

With the hardware now in place, God endows the pair with a rational soul (378). The curtain parts, and the first two humans take the stage.

A 500,000-year-old origin of Adam and Eve places them within *Homo heidelbergensis*, the big-brained, tool-making hominin thought to be ancestral to Neanderthals, Denisovans, and modern humans. Identifying the two original humans as *Homo heidelbergensis* allows Neanderthals and Denisovans membership in the image-bearing human club.

Kicking the can back 500,000 years appears to solve (at least mathematically) the problem of genetic diversity in modern humans. But what of Adam and Eve's neighbors, the other (nonmodified, nonhuman) members of *Homo heidelbergensis*? Descendants of Adam and Eve interbreeding with nonhumans is bestiality, a big ick for Craig. So, says Craig, either Adam and Eve and their descendants kept to themselves and kept their distance from other members of *Homo heidelbergensis* (apparently incest is not ick) or "fallen" *Homo heidelbergensis* engaged in bestiality (378).

Craig's concern for literalism is limited to a founding pair, which lands him in trouble with literalists of all types. Craig jettisons the "fantastic" aspects of Eden—talking snake, magic trees—calling these elements "clearly false."[25]

HELIOCENTRIC CERTAINTY

The Gauger and Hössjer study made a splash. Swamidass and Craig rejoiced. Creationists, both young-earth and old-earth, celebrated. The study appeared to debunk a popular book coauthored by Dennis Venema and Scot McKnight, *Adam and the Genome*.[26] Venema is a biologist and a frequent contributor to BioLogos and the American Scientific Affiliation, organizations committed to faith and acceptance of mainstream science.

Using the population genetics tools described in this chapter, Venema determined that modern humans evolved from a population of many thousands. Modern genetic diversity, said Venema, rules out a founding population of two.[27]

Venema agreed to kick around the numbers with Richard Buggs, a UK geneticist and a proponent of intelligent design creationism, and a few other biologists. Like Gauger and Hössjer, Buggs is a fan of a (mysterious) extreme bottleneck in the human population around 500,000 years ago—leaving only two people, of course. What's more, Buggs suggests that the reduction to two occurred in a *single* generation. To avoid the problem of extinction that often follows a severe bottleneck, Buggs suggests a second dramatic scenario: Immediately following the reduction to two people, the population grew at an exponential rate.[28]

Ultimately, Venema and Buggs determined that *if* such an event occurred, it would be detectable if it happened less than 500,000 years ago. But it was far from a slam-dunk admission of error by Venema. Only one modeling approach was used, says Venema, and the model was "torture tested," at that.

Moreover, there is absolutely no evidence of a bottleneck followed by exponential growth in hominins 500,000 years ago. At that point, hominins were living throughout Africa, Asia, and Europe—you can imagine the logistical problems of such a rapid reduction to two followed by an exponential expansion.

While a "close one eye and squint with the other" approach to the data might bring theological comfort to a concordist, none of the scenarios suggested by Buggs, Swamidass, Craig, Gauger and Hössjer, or Rana would convince a geneticist outside a religious setting.

The evidence for our evolution as a population is simply overwhelming. After crunching the numbers with Buggs, Venema stands by his original data: "We can be confident that finding evidence that we were created separately from other animals or that we descend only from two people just isn't going to happen. Some ideas in science are so well-supported that it is highly unlikely that new evidence will substantially modify them, and these are among them. The sun is at the center of our solar system, humans evolved, and we evolved as a population."[29]

Genetic evidence supports a population of thousands. The methods of population genetics, trusted in multiple areas of science, support a population of thousands. The twisting and pretzeling and mental gymnastics required to get to a "maybe?" makes Simone Biles look like a Gymboree kid.

If your theology demands a literal Adam and Eve, choose from the various scenarios suggested by creationists and concordists. Although such scenarios might quiet theological angst, these scenarios are *not* science and stand far outside the consensus of evidence regarding human origins. Intellectual honesty demands this concession.

Were Adam and Eve real people? Did they exist in human history? This is not a question for science. If your theology demands an actual Adam and Eve, so be it. What they *cannot* be, however, are the literal first pair, parents of all humanity.

Both the Christian and Hebrew Scriptures have a rich tradition of parables—stories told to communicate deep spiritual truths. No one argues for a literal good Samaritan or a literal prodigal son, yet these parables are accepted as conduits for timeless truths.

Truth, but not history. Maybe there's another way to read the Genesis story.

DISCUSSION PROMPTS

1. Who is Mitochondrial Eve? How does Mitochondrial Eve differ from Eve in the Genesis story?

2. Small, isolated populations have little genetic diversity. Why is this so?
3. The genetic methods that indicate humans evolved in a population of thousands are the same science methods used and trusted in multiple fields of modern science. Is it reasonable to trust the genetic science in areas of medicine and environmental studies but distrust it when applied to human evolution? Why or why not?
4. Describe the DNA "front-loading" solution proposed by many creationists.
5. Populations with little genetic diversity are at risk of extinction. Explain.
6. Population bottlenecks occur when a population's size is reduced dramatically for at least one generation. Common causes of bottlenecks are disease, environmental disasters, and overhunting. Why do bottleneck events threaten the future existence of a population?
7. What is concordism? Give an example. Have you heard concordist arguments other than those mentioned in this chapter?
8. Define the term "hypothesis" as used in science. How does the scientific definition of "hypothesis" differ from the common use of the term? What is your opinion of Swamidass's and Craig's efforts to concord a literal Adam and Eve with science evidence?
9. The preponderance of evidence indicates that humans evolved as a population. Alternative explanations have been proposed but are seldom found outside religious settings. Why do you think this is so?

10. Can a story be "truth" without being historically or factually "true"? Can you think of a nonbiblical story (or narrative of any sort) that is true but not factually true as twenty-first-century people define "true"?

11

MADE IN THE IMAGE

Settle in, it's time for a story. It's literally a tale of cosmic proportions.

"When on high," the story begins—that's the second millennium BCE Babylonian way of saying "once upon a time." When the curtains open, there is nothing but darkness and water. Tiamat, the goddess of the sea, is the very embodiment of dark, cold, deep chaos . . . and she is centerstage.

Tiamat and her consort produce lots of offspring and fill the cosmos with countless little gods who seldom get along. Family reunions are always fractious—think *The Maury Show* with lots of mayhem and cussing and flying chairs, except with flying swords and arrows and monsters and dragons.

Marduk, the sun god, is Tiamat's great-grandson, more powerful than any other god. Marduk is clothed in splendor and is surrounded by a bright and dazzling aura of light. Several contentious family reunions later, Marduk declares war on his great-grandmother Tiamat. Marduk defeats Tiamat and, in one violent swoop, slices her body in half.

With one-half of Tiamat's body, Marduk creates the heavens. With the other half, he creates the earth. Marduk then creates stars and arranges them in patterns, and, finally, he hangs the moon in the sky as a way to mark time.

Now the supreme leader of all, Marduk calls the subordinate gods together for a meeting. Running an entire cosmos is simply exhausting, says Marduk and his council, and they need help. So out of the bodies of previously slain gods, Marduk creates the first man, and the man and his descendants serve the gods as slaves.

Meet the humans: created by the gods to do their dirty work.

Seven hundred years later, remnants of Babylonian captivity returned to their homeland in Judah and wrote their own origin story. We read that story in Genesis—the first book of the Bible.

BREATHING THE SAME AIR

Why is the Bible's story of creation so like the Babylonian story? Clearly the two origin stories breathe the same cultural air.

In both stories, matter already exists when God (or the gods) step in and bring order out of chaos. The Hebrew word in Genesis for the chaotic "deep" is linguistically related to the name "Tiamat," the chaotic sea goddess.

In both stories, darkness precedes creation. In both stories, taming the chaos results in the separation of the heavens and the earth. In both stories, there is a firmament—a solid dome above the earth with water above it and below it.

There are differences, of course, but similarities can't be ignored.

The Babylonian creation story, Enuma Elish ("when on high"), dates to the second millennium BCE, coinciding with Marduk's popularity in the Babylonian pantheon. Most likely the story is even older, dating back to a third millennium BCE Sumerian story.[1] Whether Babylonian or Sumerian, the story far predates the Genesis version. In fact, the Hebrew in which Genesis is written did not exist until the first millennium BCE.[2]

In a prescientific world, questions of "how" and "when" were not important—hence the "once upon a time"-style opening of both Enuma Elish *and* Genesis. *Meaning* was important, not facts or a timeline.

Ancient Israel shared geographic space with the other people in Mesopotamia. Israel and her neighbors shared similar languages and practices and lifestyles. First millennium BCE exiles literally *lived* in Babylon. Israel shared a culture and a worldview with her neighbors. Is it any surprise that the "in the beginning" stories reflect a common understanding of the cosmos?

THE REBUTTAL

The first five books of the Bible, Genesis included, did not emerge from a whole cloth, written by one man (Moses) in the second millennium BCE. The overwhelming majority of biblical scholars agree: Genesis, along with the other four books, came to us through written and oral records—curated, collated, and edited—a process extending past the return of the Babylonian exiles.[3]

The finishing touches to Genesis were added by a

defeated and disillusioned people. After seventy years in Babylon, the exiles returned to Judah—a Judah with no temple, no throne, and certainly no son of David on a throne. Even the promise to Abraham seemed dashed and discarded.

Generations had grown up in Babylon. Initially, the exiles (like Daniel and his friends) kept the faith—prayed openly, refused the king's foods, and kept to themselves. The younger generations assimilated: Esther hid her religion (initially at least), married the Persian king with her guardian's blessing, and threw the king a party with all his favorite foods.

The story of Genesis was not written for the sake of simply recording history. Genesis answers the big question, the big elephant-in-the-room question: After all of this, are we still the people of God?[4]

The importance of Genesis is not as a history in the way a twenty-first-century reader understands history but is instead a *polemic*, a story intended to be a rebuttal, an opposing perspective on the status quo. Reading a polemic story as a historical narrative would be like reading Jonathan Swift's "A Modest Proposal" as a recipe book.[5]

Within the context of a shared understanding of the cosmos, how is Israel's God different from the gods of Mesopotamia? How does Genesis stand as a polemic to the earlier creation stories?

The gods of Babylon bring order to the cosmos through violence. The chaos is violent and must be tamed with violence. Order is brought about with the sword. The heavens and the earth are created via a murderous family feud and the remains of a bloody carcass.

In the Genesis story, there is also chaos. The word translated as the "deep" is linguistically related to Tiamat, the Babylonian goddess of the deep and chaotic sea. But in Genesis, God tames the chaos with words, not violence. God speaks to the chaos, and chaos responds. In Genesis, creation is good and ordered. In fact, God's "Let there be" is more of an exhortation than an order. There's no force or coercion—creation is a willing participant.[6]

And! Plot twist! In Genesis, humans aren't created to be slaves to lazy, leisure-loving gods. Humans aren't lowly worms, good only for the dirty work the gods don't want to do. To the contrary, humans are created to co-reign with God in creation.

Here's the message of Genesis: From the beginning, you have great dignity. Before Abraham and before the other patriarchs, you had dignity. Before David and before the prophets and before Babylon, you had dignity. You were created for this task. This vocation. This calling.

THE TEMPLE

We also find echoes of Genesis in the story of the Canaanite god Baal. Baal was worshiped for several aspects, including fertility, rain, and storms, but he was also the king of the Ugaritic gods. Baal assumed the role of supreme god following his defeat and subduing of Yam, the sea god—sound familiar? After the defeat of Yam, a royal temple is built for Baal in seven days. And following the seven-day project? You guessed it: Baal rested.[7]

After God defeats chaos in Genesis, no temple is built. The writer of Isaiah quotes God, saying: "Heaven is my throne, and the earth is my footstool."[8] In other words, all creation is God's temple. And when the physical tabernacle and temple were eventually built, elements of creation echoed throughout: the lampstand (tree of life), the curtains (the overhead sky), and the bronze water basin (the calm and subdued sea).[9]

In his *Lost World* series, Old Testament scholar John Walton methodically makes the case for creation as a "cosmic temple" in which God rests and reigns.[10] Paralleling ancient Near Eastern (ANE) lore, God brings order to the physical world and then sets up residence (rests) in his temple, in the span of seven days.

If the point of the Genesis narrative is ordering the chaos and establishing a cosmic temple for God (as Walton argues), elements of the story fall into place. If the goal of the narrative is *not* a historical and scientific account, seven twenty-four-hour days (as the Hebrew text implies) are no longer problematic.[11] Seven days are simply the ANE framework for the enthronement of a deity.[12] If Genesis is read with the eyes of an ANE reader, God "resting" on the seventh day is not puzzling. "Rest" in the context of an ANE deity does not mean withdrawing and relaxing; it means entering the temple, taking command, and assuming the throne of control.[13]

The original readers/hearers of the Genesis story would no doubt recognize the culturally influenced themes of temple building: God tamed the chaos, created the cosmos, and took up residence in his cosmic temple.

WHAT ABOUT THE HUMANS?

Where do humans belong in this framework? Do humans have a role in the cosmic temple?

In the ancient Near East, every major deity had a temple, and placed in the temple was a statue, or image, of the deity. After the image was carved, an elaborate "mouth opening" ritual was performed. "Mouth opening" invited the spirit of the god to inhabit the image, and the image became "alive."[14] In Genesis, God formed man from dust, and when God breathed into the man, the man became a living person.[15]

It's hard to dismiss the similarities. In fact, the Hebrew word for idol (image) is the same word used for humans in Genesis.[16] And surely, the original hearers/readers of Genesis would recognize the life-giving spirit entering the man via the breath of God.

Are humans simply the physical representation of the unseen God, made of flesh and blood instead of wood and stone? Are we only created to stand idly (or idol-y) by?

There is yet another context for "image of God" well known to the original hearers/readers of Genesis. Within this context we get a fuller understanding of what it means to be the image of God.

In the ancient Near East as well as in ancient Egypt, kings were regarded as the physical embodiment of an unseen god.[17] Kings regularly placed images of themselves across their kingdom, in the far-flung reaches of the empire, and along important borders. These images not only represented the rule of the king in these places, but also the divine right of the king to rule. It was the refusal of Shadrach, Meshach, and Abednego to worship the king's image that got them tossed into the archetypal fiery furnace.[18]

As the image of the Supreme God, King of the cosmos, humans are royalty in residence. As vice-regents, humans are given the kingly tasks of maintaining order in the realm: naming, subduing, and ruling the animals, tending the earth and watching over it.[19] Psalms picks up the royal theme: humans are crowned with glory and are made to be rulers over the works of God's hands.[20]

In the ancient Near East, the task of representing a god in the world was given to kings, but also to priests. The Levitical tasks we find assigned to priests in the Hebrew Bible are tasks also assigned to the humans in the creation story.[21] Eden, says John Walton, is sacred space, and humans are given a sacred responsibility to care for it. What's more, priestly caretaking is not simply maintaining status quo. The language in Genesis indicates that humans are to "complete what was unfinished," a continuing work of improvement, working for the flourishing of all creation.[22]

In Genesis, humans are created yet are tasked with continuing God's creative work. After all, that's how God created in the beginning—cooperatively and in concert with creation.[23] God invites active participation by the *created* in furthering creation: Let the earth bring forth grass, and fruit, and seeds. Let the waters abound with living things, and let the earth bring forth living creatures of all kinds.

LIKE PARENT, LIKE CHILD?

In the sixteenth century, Michelangelo told us that God is an old white guy in robes, a long beard, and oversized hands, reaching down from heaven to create a very buff Adam. In the eighteenth century, William Blake's *The Ancient of Days* kept the long white beard, shed the robes, and gave God a compass for calculating creation. In modern American film, God looks like either George Burns or Morgan Freeman.

Without a doubt, we've historically assumed that God looks like us, just more magisterial or, in the modern movies, funnier. We're made in God's image, so of course God looks like us, at least generally. The thought of humans evolving from monkeys, or worse, fish, or even worse, a nebulous cell in a primordial pond, is an outrage.

No one does animal humor like Gary Larson and his iconic comic *The Far Side*. Here's one of my favorites: two cows, comfy on the sofa in front of the television. On a side table, a phone rings—the old-fashioned landline kind. Cow on the right turns to her friend and says, "Well. There it goes again. And here we sit without opposable thumbs."

What exactly is the *image*, the indelible stamp that marks us and separates us from the rest of creation? What makes us more special than any other organism on the planet? Is it our opposable thumbs? Opposable thumbs certainly gave our ancestors a leg up (so to speak) in tool making, and complex tools advanced agriculture and construction.

Is it our big brains? Our big brains allowed us to develop language, and language led to cooperation and the emergence of culture. Our big brains made music, art, and algebra possible. Is it our sensory capabilities? Is it our neurological wiring that allows us to feel, react, and fear?

Because God is a spirit, is the image likewise an ethereal aspect of our being? Is the image a soul? Because God is rational, is the image our mind? Because God is the supreme ruler, is the image found in the dominance of humans over the rest of the created world? Because God is good, is the image human morality?[24]

Old Testament scholars Richard Middleton and John Walton both agree—whatever the image is, it isn't any of these things. The image of God in the Bible is not a physical or mental quality of any kind. It is not a likeness to God in any bodily sense.[25]

The image of God is not an attribute. The image of God is a function.

Clearly, our uniquely human traits play a role in the story, but these traits do not define the *image*. Both Walton and Middleton concur: The image language used in the Bible implies a task. A vocation. An assignment, if you will.[26] The assignment includes:

- Functioning as kings and priests in creation. Humans are the physical embodiment of the unphysical God. This task is both pragmatic and spiritual. As representatives of the king, humans are responsible for the flourishing of the creation-kingdom. As priests, the caretaking role is not simply physical—it is a spiritual responsibility.
- Assuming an identity. In the ancient Near East, naming is an act of creation.[27] When God names humans as image bearers, that is what we are. By naming us, it becomes us.
- Standing in for the unseen God. Humans are to be constant reminders of the qualities and character of an invisible God.

Once again, Genesis stands as a rebuttal to common cultural beliefs. Images of wood and stone are not the image of God. Kings don't own image-bearing status. Priests don't own it. To ancient Near Eastern people, this was a groundbreaking, earth-shattering concept.

Humans are the image of God. What's more, image bearing is a corporate vocation, given to all: any gender, any age, anyone along any descriptive spectrum. The geniuses among us and those with intellectual disabilities. The citizen and the immigrant. Our family, our friends, our enemies.

What's more, image bearing is a corporate vocation, given to all: any gender, any age, anyone along any descriptive spectrum.

All are given the dignity of being "specially gifted creatures of God."[28]

ADAM THE SINNER

Why is a literal Adam so important to many people of faith? For creationists like Ken Ham, it's a simple matter of authority: "If Genesis is not true, then why is the rest of the Bible true?"[29] A literal Genesis, says Ham, is the "foundation for all doctrine."[30]

For others, the desire for a literal Adam is a bit more complicated. Ideas about sin and the "fall" complicate things, and for these folks, the root of the problem is theology—the science is secondary.

Original sin is not a big deal in the tradition in which I was raised. For our group, it was more about the "age of accountability," that nebulous, unquantified year of life in which we officially become a sinner. Age of accountability notwithstanding, Adam still gets the rap as first sinner. Without a literal Adam, how could sin enter the world? And if there's no sin, there's no need for Jesus, right?

By no means. The story of Adam is the story of us all. Adam, says New Testament scholar Scot McKnight, is the "archetypal disobedient human" and reveals universal truths about human nature.[31] Paul never blames Adam for death, says McKnight: "death came to all people, because all sinned."[32]

Furthermore, Jesus never centers Adam in his teachings about the kingdom of God, nor does John or the writer of Hebrews.[33] And although Paul contrasts Jesus with Adam in his discussion of sin, says McKnight, this does not assume a historical Adam. Paul compares Jesus to Adam as a literary figure, just as the writer of Hebrews compares Jesus to another literary figure, Melchizedek.[34]

RED IN TOOTH AND CLAW

For as long as there have been humans, humans had the opportunity to do good, but they chose to do otherwise. It's pretty simple, actually. It comes with being a big-brained, intelligent, social creature. We make choices, some good, some bad, some neutral. Our instinct is to survive. Our instinct is self-gratification. Our instinct is to put ourselves and our in-group first.

In a poem exploring his grief following the death of a friend, Alfred Tennyson famously described nature as "red in tooth and claw."[35] Without a doubt, nature is a struggle for survival. The weak are not successful; the strong and ruthless are.

Biologically, success is defined simply: acquire more resources (food, shelter, mates) than others in your community. The most successful leave more offspring than the less successful. "Do to others as you would have them do to you" doesn't increase a species' success. Biologically, it is truly a dog-eat-dog world.

Humans are deeply rooted in the unfolding story of life on Earth. We share ancestry with all life—that is fact. Given this fact, how do we rise above our evolutionary instincts to put ourselves and our group first? How do we overcome our instincts to reject and dehumanize outsiders? As creatures who have been uniquely tasked with image bearing, how do we work for the flourishing of all creation?

IMAGE BEARERS

The image of God is given corporately; therefore, it cannot be God-awareness or even consciousness. The image of God is not physical or material or mental, so our evolutionary origins just don't matter. Regardless of how we got here, we are here now, and we've been given a task.

But what about the Neanderthals? What about the Denisovans? These hominins were big-brained. They were also social—so much so that we modern humans carry their DNA in our genomes. And why stop there? What about *Homo heidelbergensis*? What about Lucy?

A more important question is this: Why does it matter? If archaic human ancestors had the capacity to image God, as loved creatures (God loves ALL creation), why would they not? And if they had the capacity to image God, does that demean *our* task, as modern humans, to do the same?

Image bearing is a task, a priestly vocation. Humans

(or human ancestors) assumed the assignment when we were capable of doing so. The question of *when* just isn't important. Neanderthals? *Homo erectus*? It really doesn't matter. Worry about yourself.

DISCUSSION PROMPTS

1. Before you read this chapter, were you familiar with the Babylonian creation story? What were your initial impressions when you first heard the story?
2. Why might the Babylonian exiles question their status as the people of God?
3. Modern readers of Genesis are often puzzled by God's "resting" on the seventh day, but ancient Near Eastern readers would understand the concept. In ANE thought, what happens when a god "rests"?
4. What does "created in the image of God" mean to you? What informed your thinking—was it church, family, friends, your own explorations, or something else?
5. Read the story of the fiery furnace in the Hebrew Scriptures, Daniel 3:1–30. What does this story indicate about "the image of god" in ancient Near Eastern thought?
6. Humans are given kingly tasks in the Genesis story. What responsibilities does that phrase imply? What might that look like in a modern world?
7. Humans are given priestly tasks in creation. What responsibilities does that phrase imply? What might that look like in a modern world?

8. Describe your understanding of "original sin." What informed your thinking—was it church, family, friends, your own explorations, or something else?
9. How is nature "red in tooth and claw"? Are humans likewise disposed to such behaviors? How so?
10. "The image of God is given corporately"—agree or disagree? Explain.

12

FISH IN THE FAMILY TREE—AND THAT'S OK

I'm married to a dermatologist, so I can attest to this fact: There's no magic cream that will give you thick skin.

A few years ago, a Twitter (now X) guy posted a screenshot of my publisher's catalogue, featuring my first two books. With lots of gnashing of teeth, Twitter Guy bemoans the current state of Christian publishing and dismisses my books as "junk." One small problem—the second book had not yet been released, and it's also apparent he hadn't read the first one. My response: You dismissed the books as "junk" without reading?

Well. Twitter Guy throws down the "pig-in-need-of-a-makeover" trump card. Twitter Guy invites me to put lipstick on a pig, and not only lipstick, but also perfume, jewelry, and practically an entire Sephora if I like, but my piggies (my books apparently) will still be pigs.

And then, Twitter Guy mic drop: I don't need to read your books to know they're junk. Thick skin! Twitter Guy is a writer at one of the big young-earth creationist organizations, so I can only imagine how much he'd dislike the books if he'd actually read them.

No one hates a book as much as someone who hasn't read it. And in the seventeenth century, no one was as certain as those who refused to look through the telescope.

Galileo didn't invent the telescope, but he had the best one of his day. Galileo turned his fancy telescope toward Jupiter, and what do you know—he saw four moons orbiting the giant planet. We now know there are at least ninety-five moons orbiting Jupiter, so why the excitement over four?

In the seventeenth century, most people thought that the earth was the center of it all, and everything literally revolved around us. Moons orbiting Jupiter made headlines. Some folks, however, were harder to convince. Most had theological reasons for placing the earth at the center of the universe; if the earth isn't the center of it all, then mankind is not the center of God's attention. And just like that—Christianity falls apart.

Galileo set his telescope up in public and invited skeptics to look for themselves. Some looked but denied what they saw; the spyglass is obviously bewitched! Others, however, simply refused to look through the telescope. Hard pass. Nothing to see here.[1]

As I write, two American astronauts have just splashed down after being stranded for more than nine months aboard the International Space Station. From his bully pulpit 250 miles above Earth, astronaut Butch Wilmore recorded a Christian message. The sermon was short and standard fare: sin, Jesus, salvation.

Wilmore closed his mini-sermon with a reflection on space, galaxies, and stars, then segued to DNA and various aspects of the human body. "It's unbelievable," concluded Wilmore, "the thought that people can believe in evolution—I'm sorry, but that's faith right there, to believe that all of this came by happenstance."[2]

Wilmore is not the first person to suggest that evolution is a matter of faith, not facts, but he's probably the first to say so while in low earth orbit. In other words, there is no empirical evidence for evolution. In other words, the decades of research and the volumes of data are fiction.

Human origins can be explained by physical and material processes. It does us no good to pretend otherwise. The earth isn't flat, gravity is real, and humans evolved. The evidence is irrefutable, but will we look?

STACKING THE DECK

What most people who reject evolution know about evolution comes from *anti*-evolution apologetics resources. Apologetics starts with foregone conclusions, then cherry-picks through the evidence for answers fitting a narrow interpretation of the biblical narrative. Science, on the other hand, starts with questions and looks for answers.

Apologetic arguments lean heavily on the unexplained. When we can't explain (or comprehend) a process or a step in a process, we pop God into the gap. When we are incredulous, God is an easy answer . . . until we find the science answer. If you're hanging on to that *one* thing science hasn't yet explained to demonstrate the magnitude of God or the necessity of God, I'm afraid you'll either be disappointed or you'll have to look away.

"But what about the fine-tuning of the universe?" counters the apologist. With the tiniest of deviations in the physical constants (like gravity and electromagnetic properties), there's no universe, no solar system, no planet, no us. The anthropic principle is a belief that says the universe is incredibly precise and designed for life—specifically humans, the crown jewel of creation. In other words, the cosmic deck was stacked in our favor.

Like a historical Adam, the anthropic principle is not a question for science. Science only deals with questions of when and how, not questions of who and why. If fine-tuning with an anthropic goal is a must for your theology, that's fine; just understand it's not science. Natural processes don't have "goals."

But consider: The universe is what we have because we are here. A universe more or less than it is, and there are no humans (as we know them) around to contemplate such questions.

Perhaps a timeless God exists through oscillating universes or multiverses or expanding and collapsing universes or universes we have yet to hypothesize, until all those mathematical probabilities are solved, and finally, we're here. No big deal. It's not like God had one shot at it, and, boy, was he lucky.

NOT-SO-SPECIAL HUMANS

Research for this book included crowdsourcing from a large discussion group dedicated to issues of science and religion. Participants fall across a broad spectrum of religious faith and acceptance of evolution. I kept it simple: What are (or were) your biggest questions or hesitations regarding human evolution?

I have pages of responses, but common themes emerged. Many struggled with the implications of a purely physical origin without supernatural intervention and what that means for the status of humans in the big picture. Are we just another animal, different only in degree from a clam, a fish, or a chimpanzee?

The natural history of our planet seems to indicate that we weren't the goal, or even the priority.

The universe was already about nine billion-ish years old when the earth and the rest of our solar system formed, 4.6 billion years ago. It's hard to comprehend billions, so let's compress Earth's history into a calendar year.

The sun, Earth, and planets form on New Year's Day.

I hope you brought a book, because not much happens (life-wise) until November 18. That's when the first animals with hard parts (like shells) appear. The first four-limbed animals (descendants of animals like *Tiktaalik*) show up on December 1. The nonbird dinosaurs die in a mass extinction the day after Christmas, December 26. Finally, at 11:35 p.m. on December 31, humans make their appearance on Earth. If you prefer your Earth history as a twenty-four-hour clock, humans show up less than two seconds before midnight.

Billions of years of natural history and not a single human in sight. You might think humans were an afterthought.

Jim Stump is a philosopher and a science communicator. Stump's book, *The Sacred Chain*, is a deep dive into what it means to be human in the context of evolution. Stump acknowledges the challenge of the timeline: "Taking so much time to create seems to suggest that we were not God's top priority."[3]

Could God have fabricated an Adam out of the dust in an instant? Perhaps, but biblical miracles, says Stump, were never labor-saving devices. Maybe God enjoys the work of creating, loves it even.[4]

A baby is born following nine months of gestation—does the wait make the child less special, less miraculous? Do our physical origins, playing out over eons of time, rob of us value? Is a miracle somehow more holy? Is an instantaneous creation more noble than a God of love and wisdom who wills primal forms capable of adapting, changing, evolving?

Is an instantaneous creation more noble than a God of love and wisdom who wills primal forms capable of adapting, changing, evolving?

It is not a special act of creation that makes us worthy. Our value lies in who chose us.

MAINTAINING CREDIBILITY

Explaining Life Through Evolution by Prosanta Chakrabarty is one of my favorites.[5] In an interview with Ira Flatow on the podcast *Science Friday*, Chris from Cleveland calls in with a question for Dr. Chakrabarty. Chris's question is about human evolution, and Chris thinks it's all confusing—why don't we just go to the Bible, specifically Genesis? Genesis has all the answers and is clear, says Chris.

I have no idea what Chakrabarty's religious background is or if he practices a religion at all. Chakrabarty is polite to the caller, but he really doesn't engage. "I don't want to change anybody's minds about that," he says.

"I think it's best to let people believe how they want to believe."[6]

Let's pat the little Christian on the head and move on; we've got some science to discuss.

Is that the wise approach? I've decided it's not the best approach for me. I'm not ready to give up. I'm not yet ready to write off people of faith.

Am I fighting a battle that should not be fought? Or a battle that can't be won? I don't think so.

The Barna Group tracks church trends. In a 2023 report, Barna surveyed members of six different demographics, including practicing Christians, nonpracticing Christians, people from non-Christian faiths, and people who do not identify with any religious practice or faith. Each group was asked this question: What would cause you to doubt the claims of Christianity?[7]

The answers were fascinating, but I'd like to focus on one group—those who do not identify with any religious faith. These folks aren't necessarily atheists or agnostics. They might be the "spiritual but not religious," they might be the "nones," they might be the "unchurched," they might be "deconstructing," or they may be those who at one time were religious but are no longer.

What would make a person in the "no religion" group doubt the claims of Christianity? Second only to "the hypocrisy of religious people" was this: science. Thirty-one percent of those in this group said "science" would make them doubt the claims of Christianity.

I can't hear what you're saying about Jesus or faith or human flourishing—your denial is too loud.

When people of faith reject science, we lose credibility. We're just an isolated group who thinks the world sits on the back of a turtle.

FREE PASS

Ian McConnell is a young singer-songwriter who has been called a "Kroger-brand Ed Sheeran," and he's funny and self-deprecating enough to own it. Probably his best-known work is a catchy pop number titled "Important," a song described by one Reddit poster as "optimistic nihilism." YouTube it—the juxtaposition of upbeat music and McConnell's lyrics need to be experienced.[8] Here's the first verse:

> I'm pretty sure that life doesn't have a meaning
> And if there's a god then he doesn't look like me
> And I'm just a member of the current apex species
> But there will be another when the humans
>   go extinct

Other clever verses, interspersed with this upbeat chorus:

> I'm not important, and neither are you
> So let's do whatever we wanna do
> Bask in our cosmic insignificance
> Soak up this blip we're livin' in
> 'Cause nothing matters anyway
> Isn't that great?

I first heard the song in church as an intro to a sermon. The response of the congregants was expected—a gasp-y "tsk-tsk, how sad so many people believe that." I admit—I liked it. Loved it even. I identify with the strugglers, the questioners, the deconstructioners, the reconstructioners. Within the jarring lyrics, there's a question (or more) that demands an answer.

Are we simply the "current apex species," doomed to eventual extinction like the other 99.9 percent of all species ever to exist? Have we deluded ourselves with aspirations of a pinnacle existence, when in fact, our lives have no meaning?

Are we, as unimportant blips on the screen of life, free to do whatever we want to do? Victorious Scopes attorney William Jennings Bryan predicted it: "Equate humans with other animals as product of purely natural processes and they will act like apes."[9]

The "Important" song is not alone in its empty, godless outlook on existence. We can deny the science, or we can concede the narrative, or we can confront the nihilism head-on. What are people of faith to do?

Physical origins don't give us a free pass. Evolution doesn't let us off the hook. We are image bearers, and we have a job to do.

MY TESTIMONY

Few philosophers of religion hold my attention like Luke Janssen. Dr. Janssen is by training and profession a physiologist, but his reflections on faith as a man of science are gold. Warning: He doesn't take the easy, tie-it-all-up in a sweet morning devotional approach to God.

Love is a motivator, says Janssen. Love compels sentient beings to join as couples, families, communities, and societies.

On the quantum end of things, we also find compelling forces; small particles come together to form larger particles, then larger particles combine as atoms and simple molecules, and finally, complex molecules.

Natural forces—both chemical and physical—self-organize complex molecules into stars, planets, and living organisms.

Who's to say, asks Janssen, "that all of this attraction and building, cooperation, sharing, that these aren't the manifestations of a force called love." Perhaps, says Janssen, "that's God, or at least an aspect of God."[10]

Abiogenesis is beautiful and elegant. What a mind to conceive a world with such possibilities! Evolution is likewise elegant. Not perfection, but elegant in the process. Is a micromanaging God somehow more noble than a God who wills primal forms, capable of adapting and changing?

No common ancestry or fish ancestor or monkey in my family tree changes the nature of God or my place in creation.

This is the origin we have. This is the existence we have. Instead of being something that threatens my faith, evolution makes my view of God bigger, not smaller. My understanding of God is so much larger than the "creation in a box" story I'd believed by default.

The gospel is not so fragile as to be undone by a few science facts. My faith is in Christ alone—not in a historical reading of Genesis, not in an inerrant Bible, not in creationism. My faith is not in apologetic arguments designed to prove the fraud of evolution and the existence of God.

My one piece of religious jewelry is a simple necklace with two Greek letters, ΑΩ. I wear the alpha omega as a reminder of the one who is the Alpha and Omega, beginning and end, the first and last, age to age.

God spoke, and it was so.

DISCUSSION PROMPTS

1. With a foregone conclusion, we start with an answer and look for stuff to back us up. How does this approach differ from the scientific method?
2. When we can't yet explain something in the natural world, it is tempting to attribute the unknown step or process to divine intervention by God. What problems might arise with a "God of the gaps" approach to scientific questions?
3. An American astronaut called evolution a matter of "faith" in an orbital sermon. He's not the first person to characterize evolution as something to be accepted by "faith." What does this assumption say about the nature of evolution?
4. Explain the anthropic principle in your own words. Is the anthropic principle a scientific concept? Why or why not?
5. The universe existed for billions of years before humans showed up. Were humans an afterthought? Or does God enjoy the work of creating? What do you think?
6. In your opinion, why do many scientists disengage from conversations about faith and evolution? What are the consequences of disengagement?
7. Were you surprised to learn that science is a major stumbling block for people contemplating the claims of Christianity? Why do you think this is so?
8. Listen to the "Important" song by Ian McConnell. List a few words or feelings the song evokes for you.
9. If humans are indeed fish and apes, are we free to do as we please? Why or why not?
10. What bestows value on humanity?

ACKNOWLEDGMENTS

You can always count on family for embarrassing childhood stories.

I frequently came home from kindergarten in tears, saying, “The other kids don’t like me.” My mom scheduled a parent conference, expecting a comforting alternative story from my teacher.

Instead, my teacher responded, “She’s right.” Apparently, I was a bit too bossy during group activities.

And to this day, I’m not a fan of group projects.

But writing a book is a different story. Writing is hard work, and it takes a team for an idea to travel from brain to book. I’m beyond grateful for mine.

I owe big hugs and thanks to many:

To my husband, Mark: my books wouldn’t exist if not for your encouragement and confidence in my writing and my voice. You’re my science consultant and sounding board, and you’re the best road trip partner for scientific adventures.

To my dad, Kirby Kellogg, my lifelong cheerleader and a stellar example of loving your neighbors.

To my dear aunt Carol Mitchell, you are a Woman of Letters and a hero of faith. Reading the notes, anecdotes, and family lore you wrote in the margins of my rough drafts made editing fun.

To my Matters More book group—always a safe place to dive into the hard topics and drink deeply.

To Tommy and Melinda Ballard and Sheila Legan, for their sharp-eyed reading of early drafts, expertise, and corrections, and to Dede Jackson for reading final page proofs. You all are dear to me, and I owe you dinner!

To Kristine Johnson, woman of science and faith and skilled wrangler of online evolution discussion groups, for blessing me with her expertise and editing eye.

To my New Heritage Fellowship church community, where all means all. You are a breath of fresh air and an inspiration.

To the editorial and marketing team at Wm. B. Eerdmans Publishing Company—you're the best and I'm thankful for the privilege of writing for you. I'm happy to place my book in your hands.

To my favorite biology professor at Abilene Christian University, the late Dr. Archie Manis. I never got to tell you that you are a hero to me—a man of faith who stood up for science at great personal cost.

And finally, to my three mutts, Curtis, Julie, and Lucky, your frequent requests for snacks and yard walks broke up long stretches of research and writing.

NOTES

CHAPTER 1

1. Gutsick Gibbon, "84% Chimpanzee: The Nonsense Creationism of Jeffrey Tomkins," YouTube, accessed August 25, 2025, 2:16:24, https://tinyurl.com/35hsw6r6; a writeup of the analysis discussed in the video can be accessed at https://tinyurl.com/2s3vhymc.

2. Pseudogenes are discussed in chapter 8.

3. Charles Darwin, *On the Origin of Species or the Preservation of Favoured Races in the Struggle for Life* (John Murray, 1859).

4. Charles Darwin, *The Descent of Man and Selection in Relation to Sex* (John Murray, 1871).

5. Sam Kean, "The Case Against Charles Darwin," *Distillations Magazine*, August 17, 2023, https://tinyurl.com/2ratk6mu.

6. "Butler Act," RG 260: Acts of the General Assembly, Public and Private, 1790–present, 44122, Tennessee State Library and Archives, Tennessee Virtual Archive, March 21, 1925, https://tinyurl.com/4a5n9j8e.

7. Edward Hume, *Monkey Girl: Evolution, Education, Religion, and the Battle for America's Soul* (HarperCollins, 2007), xviii.

8. Hume, *Monkey Girl*, xi.

9. Hume, *Monkey Girl*, xix.

10. Hume, *Monkey Girl*, 222.

11. Gary Johnstone and Joseph McMaster, dir., *Judgment Day: Intelligent Design on Trial* (NOVA, 2007), https://tinyurl.com/mt7e9hyc.

12. Fazale Rana, "Evolution's Flawed Approach to Science," Reasons to Believe, August 8, 2018, https://tinyurl.com/3s6e5rp4.

13. H. Lyn White Miles and Christopher Marinello, "'I Didn't Evolve from No Monkey': Religious Narratives About Human Evolution in the U.S. Southeast," *Proceedings of the Annual Meeting of the Southern Anthropological Society* 40, no. 1 (2012): 14, https://tinyurl.com/4v3r99vn.

14. Eric Hovind (@erichovind), "Teaching children evolution (molecules-to-man / macro evolution) is a form of child abuse. Why? Because evolution is a lie that robs children of absolute morality." X, November 20, 2024, 6:56 a.m., https://tinyurl.com/4u855sr7.

15. "Public's Views on Human Evolution," Pew Research Center, December 30, 2013, https://tinyurl.com/55zbfv9c; "The Evolution of Pew Research Center's Survey Questions About the Origins and Development of Life on Earth," Pew Research Center, February 6, 2019, https://tinyurl.com/4wvy6ts4; Megan Brenan, "Majority Still Credits God for Humankind, but Not Creationism," Gallup, July 22, 2024, https://tinyurl.com/2fdw2zdk.

16. Michael B. Berkman and Eric Plutzer, "Defeating Creationism in the Courtroom, But Not in the Classroom," *Science* 331, no. 6016 (2011): 404–5, https://doi.org/10.1126/science.1198902.

17. Olga Khazan, "I Was Never Taught Where Humans Came From," *Atlantic*, September 19, 2019, https://tinyurl.com/cu5awevy.

18. Britteny Berumen, Misty Boatman, and Mark W. Bland, "Public vs. Private: High School Biology Teachers' Acceptance and Teaching of Evolutionary Theory in Arkansas," *American Biology Teacher* 86, no. 2 (2024): 87–93, https://doi.org/10.1525/abt.2024.86.2.87.

19. Luke Janssen, "Human Evolution, Morality, and Our Ultimate Purpose," June 28, 2024, in *Recovering Evangelicals*, podcast, host Luke Janssen, 56:45, https://tinyurl.com/nhfy88sj.

20. Prosanta Chakrabarty, *Explaining Life Through Evolution* (MIT Press, 2023), 9, fig. 1.

21. Jack Ashby, "The Absurdity of Natural History—or, Why Humans Are Fish," *Conversation*, November 29, 2016, https://tinyurl.com/3a4hen5k.

CHAPTER 2

1. Bill Bryson, *A Short History of Nearly Everything* (Broadway Books, 2003), 287.

2. The 747 analogy is attributed to Fred Hoyle in his book *The Intelligent Universe* (1983). Although Hoyle considered himself an atheist or a rationalist, the 747 analogy is a favorite of creationists.

3. Embark Ministries with Brad Small, "How Can I Know There Is a God Part 2," YouTube, August 19, 2021, 21:07, https://tinyurl.com/m45nw9xc.

4. Michael Shermer, "To Be or Not to Be a Weasel: Hamlet, Intelligent Design, and How Evolution Works," *Skeptic* 9, no. 4 (2002): 16–20.

5. Simon Conway Morris, "We Were Meant to Be . . . ," *New Scientist* 176, no. 2369 (2002): 26.

6. Carl Zimmer, "On the Origin of Life on Earth," *Science* 323, no. 5911 (2009): 198–99.

7. Wynne Parry, "Possible Key to Life's Chemistry Revealed in 50-Year-Old Experiment," *Live Science*, March 21, 2011, https://tinyurl.com/39u8yf2z.

8. Jennifer Ouellette, "Scientists Recreated Classic Origin-of-Life Experiment and Made a New Discovery," Ars Technica, October 28, 2021, https://tinyurl.com/4fxxrwk7.

9. Douglas Fox, "Primordial Soup's On: Scientists Repeat Evolution's Most Famous Experiment," *Scientific American*, March 28, 2007, https://tinyurl.com/2bv2b8ft; Hannah Hickey, "Tiny, Ancient Meteorites Suggest Early Earth's Atmosphere Was Rich in Carbon Dioxide," *UW News*, January 24, 2020, https://tinyurl.com/37zaubw9.

10. Angela M. Cowan, "Deep Sea Hydrothermal Vents," *National Geographic*, last updated October 19, 2023, https://tinyurl.com/kzptte9n.

11. Peter Biro, "Questions and Answers About Hydrothermal Vents," Marine Education Society of Australasia, accessed August 25, 2025, https://tinyurl.com/4zz2vupd; Rachel Brazil, "Hydrothermal Vents and the Origins of Life," *Chemistry World*, April 16, 2017, https://tinyurl.com/j2m868tf.

12. Henry Gee, *A (Very) Short History of Life on Earth* (St. Martin's, 2023), 4–5; Henry Gee, "The Origins of Life on Earth," December 1, 2021, in *The Ancients*, podcast, host Tristan Hughes, 1:24:00, https://tinyurl.com/4xd79d5r.

13. Eugenie C. Scott, *Evolution vs. Creationism: An Introduction* (University of California Press, 2009), 26.

14. Russ Miller, "Defending Biblical Creation" (Steeling the Mind conference, Coeur d'Alene, ID, February 16, 2019).

15. Gee, "The Origins of Life on Earth."

16. Edmund R. R. Moody et al., "The Nature of the Last Universal Common Ancestor and Its Impact on the Early Earth System," *Nature Ecology and Evolution*, July 12, 2024, https://doi.org/10.1038/s41559-024-02461-1.

17. Robert F. Service, "Our Last Common Ancestor Lived 4.2 Billion Years Ago—Perhaps Hundreds of Millions of Years Earlier Than Thought," *Science*, July 12, 2024, doi: 10.1126/science.z2x162s; Nicholas Wade, "Meet Luca, the Ancestor of All Living Things," *New York Times*, July 25, 2016, https://tinyurl.com/m34vwawe; Madeline C. Weiss et al., "The Last Universal Common Ancestor Between Ancient Earth Chemistry and the Onset of Genetics," *PLoS Genetics*, August 16, 2018, https://doi.org/10.1371/journal.pgen.1007518.

18. Gee, *A (Very) Short History of Life on Earth*, 6; Gee, "The Origins of Life on Earth."

19. Stephen Porder, *Elemental: How Five Elements Changed Earth's Past and Will Shape Our Future* (Princeton University Press, 2023), 8–9.

20. Madison Dapcevich, "Mass Extinction Event 2 Billion Years Ago Killed 99% of Life on Earth, Study Finds," EcoWatch, September 4, 2019, https://tinyurl.com/ddxr5vhb; Kartik Aiyer, "The Great Oxidation Event: How Cyanobacteria Changed Life," American Society for Microbiology, February 18, 2022, https://tinyurl.com/sn9ytt6f; "The Bacteria That Changed the World," Understanding Evolution, May 2017, https://tinyurl.com/cufsv2ws.

21. Bryan Pietsch, "A Professor Asked for 'Anti-cheating Hats.' Her Students Went All Out," *Washington Post*, October 25, 2022, https://tinyurl.com/mtdwf3pa; James FitzGerald, "Philippines: Student 'Anti-cheating' Exam Hats Go Viral," BBC, October 23, 2022, https://tinyurl.com/46zkztm6.

22. Read more about Margulis's work: "Lynn Margulis and the Origins of Multicellular Life," *More Than a Dodo*, the Oxford University Museum of Natural History blog, March 8, 2019, https://tinyurl.com/mvx9yday.

23. "Finding Our Roots," Understanding Evolution, accessed August 25, 2025, https://tinyurl.com/8eakpy99.

24. Lynn Margulis and Dorion Sagan, *Microcosmos: Four Billion Years of Evolution from Our Microbial Ancestors* (University of California Press, 1986), 29.

25. Amanda Heidt, "The Long and Winding Road to Eukaryotic Cells," *Scientist Digest*, October 16, 2022, https://tinyurl.com/28624fp4; James Han, "The Origin of the Nucleus: The Discovery of a Cellular Fossil," *Yale Scientific*, January 29, 2020, https://www.yalescientific.org/2020/01/the-origin-of-the-nucleus-the-discovery-of-a-cellular-fossil/; Florence Camu, Mike Murphy, and Nick Lane, "Mitochondria," June 29, 2023, in *In Our Time*, podcast, host Melvin Bragg, 55:09, https://tinyurl.com/pbm8sjed.

26. Donald R. Prothero, *Evolution: What the Fossils Say and Why It Matters* (Columbia University Press, 2017), 173–75.

27. Richard K. Grosberg and Richard R. Strathmann, "The Evolution of Multicellularity: A Minor Major Transition?" *Annual Review of Ecology, Evolution, and Systematics* 38 (2007): 621–54.

28. Siddhartha Mukherjee, *The Song of the Cell: An Exploration of Medicine and the New Human* (Scribner, 2022), 131.

29. Elizabeth Mitchell, "Attempts to Trace Life Back to Chemical Origins *Still* Maps the Willful Ignorance of the Hunters," Answers in Genesis, May 2, 2015, https://tinyurl.com/ywbv6s7p.

30. Frank Sherwin, "What Is the Origin of Life?," Institute for Creation Research, February 28, 2014, https://tinyurl.com/5n93nsu5; David Menton, "The Origin of Life," Answers in Genesis, June 24, 2017, https://tinyurl.com/5y2744z4; Casey Luskin, "New Study Triggers Key Origin of Life Questions," December 2, 2024, in *ID the Future*, podcast, host Andrew McDiarmid, 20:19, https://tinyurl.com/yd72mzry; Neil Thomas, "Considering 'Abiogenesis,' an Imaginary Term in Science," Science and Culture Today, April 11, 2022, https://tinyurl.com/4wjydn77.

31. Mitchell, "Attempts to Trace Life Back to Chemical Origins *Still* Maps the Willful Ignorance of the Hunters."

CHAPTER 3

1. Adrian Desmond and James Moore, *Darwin* (Warner Books, 1991), 257.

2. Read about Alfred Russel Wallace: James McNish, "Who Was Alfred Russel Wallace?" Natural History Museum, accessed August 26, 2025, https://tinyurl.com/5h2hbp4h.

3. Brian Switek, "Did the Cambrian Really Give Darwin Nightmares?" *Wired*, September 16, 2009, https://tinyurl.com/yeyny7a2 (note: Switek now writes as Riley Black).

4. "Trilobites," British Geological Survey, accessed August 26, 2025, https://tinyurl.com/4kw5cvdc.

5. "Darwin's Predictions," NOVA, accessed August 26, 2025, https://tinyurl.com/3nftfp4x.

6. Charles Darwin describes his Cambrian dilemma in chapter 9, "On the Imperfections of the Geological Record," in *On the Origin of Species or the Preservation of Favoured Races in the Struggle for Life* (John Murray, 1859).

7. Darwin, "On the Imperfections of the Geological Record."

8. A "billycan" is a small metal can used for boiling water over an open fire.

9. Kristen Weidenbach, *Rock Star: The Story of Reg Sprigg—an Outback Legend* (Midnight Sun Publishing, 2014).

10. J. William Schopf, "Solution to Darwin's Dilemma: Discovery of the Missing Precambrian Record of Life," *PNAS* 97, no. 13 (2000): 6947–53, https://doi.org/10.1073/pnas.97.13.6947.

11. Weidenbach, *Rock Star*.

12. Katherine J. Wu, "How Does Microwaving Grapes Create Plumes of Plasma?" NOVA, February 18, 2019, https://tinyurl.com/yeyu4fvn.

13. Jacob Beasecker et al., "It's Time to Defuse the Cambrian 'Explosion,'" *GSA Today* 30, no. 12 (2020): 26–27, https://doi.org/10.1130/GSATG460GW.1; "The Cambrian Period," UC Museum of Paleontology, July 6, 2011, https://tinyurl.com/3byjxbd7.

14. Thomas Servais et al., "No (Cambrian) Explosion and No (Ordovician) Event: A Single Long-Term Radiation in the Early Palaeozoic," *Palaeogeography, Palaeoclimatology, Palaeoecology* 623 (2023): 1–20, https://doi.org/10.1016/j.palaeo.2023.111592; Betsy

Mason, "Long Fuse for Cambrian Explosion," *Science*, April 13, 2004, https://tinyurl.com/3pkp7ues.

15. Nigel C. Hughes, "Creationism and the Emergence of Animals: The Original Spin," *Reports of the National Center for Science Education* 20, no. 5 (2000), https://tinyurl.com/4ufktt38; James E. Platt, "Why 'Sudden Appearance' Is Not as It Appears," *Evolution Education Outreach* 2 (2009): 676–78, https://doi.org/10.1007/s12052-009-0165-9; David Campbell and Keith Miller, "The 'Cambrian Explosion': A Challenge to Evolutionary Theory?" in *Perspectives on an Evolving Creation*, ed. Keith Miller (Eerdmans, 2003), 182–204; M. Zhu et al., "A Deep Root for the Cambrian Explosion: Implications of New Bio- and Chemostratigraphy from the Siberian Platform," *Geology* 45, no. 5 (2017): 459–462, https://doi.org/10.1130/G38865.1; Beasecker et al., "It's Time to Defuse the Cambrian 'Explosion.'"

16. Douglas Fox, "What Sparked the Cambrian Explosion?" *Scientific American*, February 16, 2016, https://tinyurl.com/3c9v7mk7.

17. Rebecca M. Flowers et al., "Diachronous Development of Great Unconformities Before Neoproterozoic Snowball Earth," *PNAS* 117, no. 19 (2020): 10172–80, https://doi.org/10.1073/pnas.1913131117.

18. Jasna Hodžić, "Biological Big Bang: How We Solved Darwin's Dilemma," Big Think, April 5, 2022, https://tinyurl.com/2x3tmtbd.

19. J. John Sepkoski, "Foundations: Life in the Oceans," in *The Book of Life*, ed. Stephen Jay Gould (Norton, 2001), 52.

20. Flowers et al., "Diachronous Development of Great Unconformities Before Neoproterozoic Snowball Earth."

21. "Homeotic Genes and Body Patterns," Genetic Science Learning Center, University of Utah, March 1, 2016, https://tinyurl.com/4vrwrsv8.

22. "Understanding Complexity," Understanding Evolution, accessed August 26, 2025, https://tinyurl.com/3a658evr.

23. Sean B. Carroll, *Endless Forms Most Beautiful: The New Science of Evo Devo and the Making of the Animal Kingdom* (Norton, 2005), 287–90.

24. Fritz Ridenour, *Who Says?* (Regal Books, 1968), 114.

25. Jeffrey P. Tomkins, "The Fossils Still Say No: The Cambrian

Explosion," Institute for Creation Research, November 30, 2020, https://tinyurl.com/3xzy7tap.

26. Jeff Miller, "The Cambrian Explosion: Falsification of Darwinian Evolution," *Reason & Revelation* 36, no. 5 (2016): 57–59, https://tinyurl.com/229mema3.

27. Dominic Statham, "The Cambrian Explosion: The Fossils Point to Creation, Not Evolution," *Creation* 39, no. 2 (2017): 20–23, https://tinyurl.com/y96nxm37; The Center for Science and Culture, "The Scientific Controversy over the Cambrian Explosion," Discovery Institute, June 12, 2019, https://tinyurl.com/uzh73hed; Miller, "The Cambrian Explosion."

28. Statham, "The Cambrian Explosion."

29. Phillip E. Johnson, *Darwin on Trial* (Regnery, 1991), 77.

30. Graham Budd and Soren Jensen, "A Critical Reappraisal of the Fossil Record of the Bilaterian Phyla," *Biological Reviews* 75 (2000): 253–95.

31. Statham, "The Cambrian Explosion."

32. Ted Allen, "Scholars Share Evidence of Biblical Flood, Call Event Relevant to This Generation," Liberty University, March 24, 2023, https://tinyurl.com/bbrsxs8c.

33. Elizabeth Mitchell, "Researchers Devise Alternate Theory for Cambrian Explosion," Answers in Genesis, December 3, 2011, https://tinyurl.com/msj7ju9u; A. Peter Galling, "Where Are All the Bunny Fossils?" Answers in Genesis, March 6, 2009, https://tinyurl.com/4u59w2jx; Elizabeth Mitchell, "Cambrian Explosion or Creation Week—Key to Vertebrate Success?" Answers in Genesis, March 12, 2015, https://tinyurl.com/3bsmvu2p; Miller, "The Cambrian Explosion"; Statham, "The Cambrian Explosion."

34. Kurt P. Wise, "One: Life's Unexpected Explosion," Answers in Genesis, January 1, 2010, https://tinyurl.com/mpc9f833.

35. Mitchell, "Cambrian Explosion or Creation Week."

36. I wrote about flood geology in *Baby Dinosaurs on the Ark? The Bible and Modern Science and the Trouble of Making It All Fit* (Eerdmans, 2021).

CHAPTER 4

1. Ashley Hamer, "99 Percent of the Earth's Species Are Extinct—but That's Not the Worst of It," Discovery, August 1, 2019, https://tinyurl.com/mtbjdtx8.

2. More about the five previous mass extinctions and a possible sixth: Michael Greshko, "What Are Mass Extinctions, and What Causes Them?" *National Geographic*, September 26, 2019, https://tinyurl.com/5d3ps5c2.

3. More about "living fossils": Sruthi Gurudev, "These Five 'Living Fossils' Still Roam the Earth," *National Geographic*, March 13, 2024, https://tinyurl.com/42wfm9j8.

4. "Building a Human Body," NPR, July 5, 2010, https://tinyurl.com/2se7wx8f; Kevin J. Peterson et al., "The Ediacaran Emergence of Bilaterians: Congruence Between the Genetic and the Geological Fossil Records," *Philosophical Transactions of the Royal Society B* 363 (2008): 1435–43, https://doi.org/10.1098/rstb.2007.2233.

5. Boyce Rensberger, "Evolution: Getting to the Gut of the Matter," *Washington Post*, July 10, 1995, https://tinyurl.com/5bdx9yud.

6. "*Pikaia*: A Chordate," Understanding Evolution, accessed August 26, 2025, https://tinyurl.com/4etjmc9.

7. "Your Inner Monkey," episode 3 in the series "Your Inner Fish," hosted by Neil Shubin (HHMI BioInteractive, 2014), https://tinyurl.com/28pvvukh.

8. "The Great Debate," Oxford University Museum of Natural History, accessed August 26, 2025, https://tinyurl.com/4e2zwrtn; Stephen Foster, "What Actually Happened at the 1860 Wilberforce-Huxley Debate?" The Victorian Web, last modified July 30, 2022, https://tinyurl.com/226p53j9.

9. Michael Mosley, "Anatomical Clues to Human Evolution from Fish," BBC News, May 5, 2011, https://tinyurl.com/ym8j86n9; Allison Wu, "Are Humans Really Descended from Fish?" *Buckingham Browne & Nichols' STEM Magazine*, February 24, 2022, https://tinyurl.com/2ykbk6np; "Face Development in the Womb," BBC, YouTube, May 3, 2011, 1:17, https://tinyurl.com/y949au4a.

10. Kellie B. Gormly, "The Curious Case of Charles Osborne, Who Hiccupped for 68 Years Straight," *Smithsonian Magazine*, June 13, 2022, https://tinyurl.com/44h346xx.

11. Neil Shubin, *Your Inner Fish: A Journey into the 3.5-Billion-Year History of the Human Body* (Pantheon, 2008), 190–92.

12. Shubin, *Your Inner Fish*, 193–95.

13. Jerry A. Coyne, *Why Evolution Is True* (Viking, 2009), 85.

14. Kate Wong, "Newfound Fossil Is Transitional Between Fish and Landlubbers," *Scientific American*, April 6, 2006, https://tinyurl.com/5yzud6av.

15. Shubin, *Your Inner Fish*, 41.

16. Jonathan Tweet, *Grandmother Fish: A Child's First Book of Evolution* (Feiwel & Friends, 2015).

17. Originally released in theaters, the film can be accessed here: Thomas Purifoy Jr., "Is Genesis History?" YouTube, March 26, 2020, 1:44:06, https://tinyurl.com/mu9xdcpu.

18. Originally aired in theaters following the feature film: Thomas Purifoy Jr., "Q&A with Del Tackett and the Scientists," Is Genesis History, accessed August 26, 2025, 14:09, https://tinyurl.com/ja64pc57.

19. Alex Brashears, "How Did Ancient Fish Make the Evolutionary Jump from Gills to Lungs?" ASU—Ask a Biologist, July 25, 2016, https://tinyurl.com/25zzzd7k.

20. Neil Shubin, *Some Assembly Required: Decoding Four Billion Years of Life, from Ancient Fossils to DNA* (Pantheon, 2020), 16.

21. Linda B. Glaser, "CT Scanner Helps Answer 150-Year-Old Question of Lung Evolution," *Cornell Chronicle*, February 6, 2013, https://tinyurl.com/mpdbz7t3.

22. Guojie Zhang and Maria Hornbek, "We're More like Primitive Fishes Than Once Believed," University of Copenhagen, Faculty of Science, February 4, 2021, https://tinyurl.com/37umyhjd; Xupeng Bi et al., "Tracing the Genetic Footprints of Vertebrate Landing in Non-Teleost Ray-Finned Fishes," *Cell* 184, no. 5 (2021): 1377–91, https://doi.org/10.1016/j.cell.2021.01.046.

23. Zhang and Hornbek, "We're More like Primitive Fishes Than Once Believed."

24. Zhang and Hornbek, "We're More like Primitive Fishes Than Once Believed."

25. Prosanta Chakrabarty, *Explaining Life Through Evolution* (MIT Press, 2023), 28.

26. Susanne Keipert et al., "Two-Stage Evolution of Mammalian Adipose Tissue Thermogenesis," *Science* 384, no. 6700 (2024): 1111–17, https://doi.org/10.1126/science.adg1947.

CHAPTER 5

1. Steve Brusatte, *The Rise and Reign of Mammals: A New History from the Shadow of the Dinosaurs to Us* (Mariner Books, 2023), 24–27.

2. Sean P. Modesto et al., "The Oldest Parareptile and the Early

Diversification of Reptiles," *Proceedings of the Royal Society B* 282, no. 1801 (2015), https://doi.org/10.1098/rspb.2014.1912; Brusatte, *The Rise and Reign of Mammals*, 5.

3. Turtles are reptiles, but they do not have the two openings in the skull found in other reptiles. Turtle DNA is definitely reptilian, so the closed turtle skull is apparently a mutation in the turtle lineage.

4. The Doubleclicks, "Dimetrodon," YouTube, October 1, 2013, 04:35, https://tinyurl.com/222by56z.

5. "The Making of Mass Extinctions," HHMI BioInteractive, accessed August 26, 2025, https://tinyurl.com/t9yd6z3a; Sarda Sahney and Michael J. Benton, "Recovery from the Most Profound Mass Extinction of All Time," *Proceedings of the Royal Society B* 275, no. 1636 (2008): 759–65, https://doi.org/10.1098/rspb.2007.1370.

6. Brusatte, *The Rise and Reign of Mammals*, 47.

7. Adam K. Huttenlocker, "Body Size Reduction in Nonmammalian Eutheriodont Therapsids (Synapsida) During the End-Permian Mass Extinction," *PLoS ONE* 9, no. 2 (2014), https://doi.org/10.1371/journal.pone.0087553.

8. Hillel J. Hoffman, "The Permian Extinction—When Life Nearly Came to an End," *National Geographic*, October 15, 2024, https://tinyurl.com/bdr39uwp.

9. Brusatte, *The Rise and Reign of Mammals*, 55–56.

10. Donald R. Prothero, *Evolution: What the Fossils Say and Why It Matters* (Columbia University Press, 2017), 296; Brusatte, *The Rise and Reign of Mammals*, 61.

11. Brusatte, *The Rise and Reign of Mammals*, 59–60.

12. Neal Anthwal, Leena Joshi, and Abigail S. Tucker, "Evolution of the Mammalian Middle Ear and Jaws: Adaptations and Novel Structures," *Journal of Anatomy* 222, no. 1 (2012): 147–60, https://doi.org/10.1111/j.1469-7580.2012.01526.x.

13. Lisa A. Urry et al., *Campbell Biology* (Pearson, 2017), 529.

14. Brusatte, *The Rise and Reign of Mammals*, 68.

15. Christine Janis, "Victors by Default: The Mammalian Succession," in *The Book of Life*, ed. Stephen Jay Gould (Norton, 2001), 172; Brusatte, *The Rise and Reign of Mammals*, 104.

16. "The Day the Mesozoic Died," HHMI BioInteractive, updated May 14, 2020, https://tinyurl.com/2wp3bkfb.

17. In multiple locations around the world, there is a thin layer

of clay separating the rocks of the Cretaceous period (right before the extinction) and the rocks of the Tertiary period (right after the extinction). This thin clay layer is rich in iridium, an element rare on Earth but common in meteorites and asteroids.

18. "Out of the Ashes: Dawn of the Age of Mammals," HHMI BioInteractive, updated August 18, 2021, https://tinyurl.com/y26t8zrt.

19. "Out of the Ashes."

20. Coevolution is a kind of "arms race." There is a moth in Madagascar with an eleven-inch-long tongue. Growing in the same niche as the moth is an orchid with an eleven-inch-long nectar tube. If a moth easily drinks nectar without pressing its head against the pollen-laden flower, pollen will not disperse and the orchid will not reproduce. Natural selection therefore favors orchids with long nectar tubes—moths lean into drink nectar, have their fill, and fly off covered in pollen. Natural selection also favors moths with long tongues able to reach the nectar. As tubes got longer, tongues got longer.

21. Brian K. Hall, "The Paradoxical Platypus," *Bioscience* 49, no. 3 (1999): 211–18, https://doi.org/10.2307/1313511; Brusatte, *The Rise and Reign of Mammals*, 144–48.

22. Abby Ohlheiser, "The Platypus Is So Weird That Scientists Thought the First Specimen Was a Hoax," *Washington Post*, April 1, 2025, https://tinyurl.com/46vx5h3v.

23. Hall, "The Paradoxical Platypus."

24. Duane T. Gish, *Evolution: The Fossils Still Say No!* (Institute for Creation Research, 1985), 40–42, 147–208.

25. Duane T. Gish, "Creation, Evolution, and the Historical Evidence," *American Biology Teacher* 34, no. 3 (1973): 132–40.

26. Elizabeth Mitchell, "Review: Your Inner Reptile," Answers in Genesis, April 19, 2014, https://tinyurl.com/3rmxjcc6.

27. Prothero, *Evolution*, 296–301.

CHAPTER 6

1. Mark Maslin, "How a Changing Landscape and Climate Shaped Early Humans," *Conversation*, November 7, 2013, https://tinyurl.com/527whn3d.

2. Maslin, "How a Changing Landscape and Climate Shaped Early Humans."

3. Mary Caperton Morton, "One Whale's Incredible Journey Details East Africa's Uplift," *Earth*, June 16, 2015, https://tinyurl.com/2etz2r5w; Mihai Andrei, "Beaked Whale Reveals Africa's Tectonic Secrets," ZME Science, March 18, 2015, https://tinyurl.com/y3bxstnf; Henry Wichura et al., *PNAS* 112, no. 13 (2015): 3910–15, https://doi.org/10.1073/pnas.1421502112.

4. Yves Coppens, "East Side Story: The Origin of Humankind," *Scientific American*, May 1994, 86–95.

5. Geoffrey Mohan, "Fossilized Whale Bone in African Desert Holds Clues to Human Evolution," *Los Angeles Times*, March 16, 2015, https://tinyurl.com/3fyyfr3b.

6. Steve Brusatte, *The Rise and Reign of Mammals: A New History from the Shadow of the Dinosaurs to Us* (Mariner Books, 2023), 367–70; Robert Sanders, "Our Earliest Primate Ancestors Rapidly Spread After Dinosaur Extinction," UC Berkeley Research, February 24, 2021, https://tinyurl.com/4mw29tdy.

7. Amanda Fiegl, "World's Oldest Primate Was a Rodentlike Climber," *National Geographic*, October 26, 2012, https://tinyurl.com/bdfbp8dt; Brusatte, *The Rise and Reign of Mammals*, 369.

8. Jim Shelton, "Fossil Ankles Indicate Earth's Earliest Primates Lived in Trees," *YaleNews*, January 19, 2015, https://tinyurl.com/ycxb6ezn; Fiegl, "World's Oldest Primate Was a Rodentlike Climber."

9. Michael O. Woodburne, Gregg F. Gunnell, and Richard K. Stucky, "Climate Directly Influences Eocene Mammal Faunal Dynamics in North America," *PNAS* 106, no. 32 (2009): 13399–403, https://doi.org/10.1073/pnas.0906802106; A. Sluijs et al., "The Palaeocene-Eocene Thermal Maximum Super Greenhouse: Biotic and Geochemical Signatures, Age Models and Mechanisms of Global Change," in *Deep-Time Perspective on Climate Change: Marrying the Signal from Computer Models and Biological Proxies*, ed. M. Williams et al. (Geological Society of London, 2007), 323–49.

10. Brusatte, *The Rise and Reign of Mammals*, 216–17.

11. Brusatte, *The Rise and Reign of Mammals*, 371–72.

12. Brusatte, *The Rise and Reign of Mammals*, 372–73.

13. Riley Black, "The Unlikely Event of Ancient Monkeys,

Swept Across the Atlantic Ocean," *National Geographic*, July 19, 2023, https://tinyurl.com/u6b5xar2; Riley Black, "More Than 30 Million Years Ago, Monkeys Rafted Across the Atlantic to South America," *Smithsonian Magazine*, April 9, 2020, https://tinyurl.com/bdcwcx7v.

14. Chris Palmer, "Fossils Indicate Common Ancestor for Two Primate Groups," *Nature*, May 15, 2013, https://doi.org/10.1038/nature.2013.12997.

15. Travis Stewart, "Consul the Great: Top Ape in Vaudeville," *Travalanche*, July 7, 2013, https://tinyurl.com/mjjb5s4c.

16. Peter Andres and Christopher Stringer, "The Primates' Progress," in *The Book of Life*, ed. Stephen Jay Gould (Norton, 2001), 221.

17. Michon Scott, "Henry Fairfield Osborn," Strange Science, updated January 7, 2024, https://tinyurl.com/5ervhes3; Benjamin Miller, "The Osborn Problem," *Extinct Monsters*, July 17, 2013, https://tinyurl.com/msemkvp8.

18. Gowan Dawson, "Helen Ziska: The Unknown Artist Who Kept the March of Progress Image Alive," Yale University Press, March 25, 2024, https://tinyurl.com/39aks8un.

19. "Walking Upright," Smithsonian National Museum of Natural History, last updated January 3, 2024, https://tinyurl.com/545z9826.

20. Beth Blaxland and Fran Dorey, "Shorter Jaws with Smaller Teeth," Australian Museum, November 9, 2018, https://tinyurl.com/8by3x5um; Herman Pontzer, "Overview of Hominin Evolution," *Nature Education Knowledge* 3, no. 10 (2012), https://tinyurl.com/ywyjzv8x.

21. Jerry A. Coyne, *Why Evolution Is True* (Viking, 2009), 203; Blaxland and Dorey, "Shorter Jaws with Smaller Teeth."

22. "Compare Lucy: Femur," eLucy, University of Texas at Austin Department of Anthropology, accessed February 28, 2025, https://elucy.org.

23. "Walking Upright."

24. "What Makes Us Special," *Scientific American* 311, no. 3 (2014): 60–61.

25. Michael Price, "Genes Reveal How Our Pelvis Evolved for Upright Walking," *Science*, August 17, 2022, https://tinyurl.com

/yj4ntmd8; Laura Tobias Gruss and Daniel Schmitt, "The Evolution of the Human Pelvis: Changing Adaptations to Bipedalism, Obstetrics and Thermoregulation," *Philosophical Transactions of the Royal Society London B* 370, no. 1663 (2015), https://doi.org/10.1098/rstb.2014.0063; Coyne, *Why Evolution Is True*, 202–4.

26. "What Makes Us Special," 60–61; "Walking Upright."

27. "Bigger Brains: Complex Brains for a Complex World," Smithsonian National Museum of Natural History, last updated January 3, 2024, https://tinyurl.com/bdzx8a9b.

28. Robin Anne Smith, "How Did Human Brains Get to Be So Big?" *Scientific American*, February 21, 2012, https://tinyurl.com/38jvte3p.

29. Sir Arthur Conan Doyle, "The Bascombe Valley Mystery," in *Sherlock Holmes: The Complete Novels and Stories* (Bantam Classics, 2003), 310.

30. Miriam Bibby, "Piltdown Man: Anatomy of a Hoax," Historic UK, accessed August 27, 2025, https://tinyurl.com/2p9zhxrx; Robin McKie, "Piltdown Man: British Archaeology's Greatest Hoax," *Guardian*, February 4, 2012, https://tinyurl.com/rupezy9w; "Piltdown Man," Natural History Museum, accessed August 27, 2025, https://tinyurl.com/2s4x2wt7.

31. "Arthur Conan Doyle Is Piltdown Suspect," *New York Times*, August 2, 1983, https://tinyurl.com/3n8ax5ue.

32. Randy J. Guliuzza, "Major Evolutionary Blunders: The Imaginary Piltdown Man," Institute for Creation Research, November 30, 2015, https://tinyurl.com/5zswd6yd.

33. Answers in Genesis, "Three Ways to Make an Apeman," YouTube, June 9, 2016, 16:06, https://tinyurl.com/mxumu6ut.

34. Erika is a primatologist who makes informative and entertaining videos as "Gutsick Gibbon." See these for more information regarding human traits and fossils: Gutsick Gibbon, "Primatology, Proof, and the Human Monkey," YouTube, February 8, 2022, 27:18, https://tinyurl.com/55ubbybr; Gutsick Gibbon, "Dear Creationists: You're Still Apes," YouTube, July 30, 2020, 1:04:11, https://tinyurl.com/yc5cwc35.

35. Brian G. Richmond and William L. Jungers, "Hominin Proximal Femur Morphology from the Tugen Hills to Flores," in *African Genesis: Perspectives on Hominin Evolution*, ed. Sally C.

Reynolds and Andrew Gallagher (Cambridge University Press, 2012), 248–67.

CHAPTER 7

1. C. K. Brain, "Raymond Dart and Our African Origins," in *A Century of Nature: Twenty-One Discoveries That Changed Science and the World*, ed. Laura Garwin and Tim Lincoln (University of Chicago Press, 2003), 3–9.

2. "What Does It Mean to Be Human?" Smithsonian National Museum of Natural History, accessed August 27, 2025, https://tinyurl.com/yp3c7d9n.

3. Jerry A. Coyne, *Why Evolution Is True* (Viking, 2009), 199.

4. Herman Pontzer, "Overview of Hominin Evolution," *Nature Education Knowledge* 3, no. 10 (2012), https://tinyurl.com/ywyjzv8x; Erin Wayman, "The Human Story," Science News, September 15, 2021, https://tinyurl.com/5e6vmkea; "*Ardipithicus ramidus*," Smithsonian National Museum of Natural History, last updated January 3, 2024, https://tinyurl.com/bdds63h8.

5. Kermit Pattison, "Lucy and Ardi: The Two Fossils That Changed Human History," BBC Science Focus, March 7, 2021, https://tinyurl.com/v72ezbfd.

6. "*Australopithecus afarensis*," Smithsonian National Museum of Natural History, last updated January 3, 2024, https://tinyurl.com/4bhx25dc.

7. Coyne, *Why Evolution Is True*, 201–2; Lisa Hendry, "*Australopithecus afarensis*, Lucy's Species," Natural History Museum, April 1, 2020, https://tinyurl.com/yk4ssbkw.

8. Coyne, *Why Evolution Is True*, 202–3.

9. Beth Blaxland and Fran Dorey, "*Paranthropus* Genus," Australian Museum, April 29, 2022, https://tinyurl.com/kh46273z.

10. Ralf Rotheimer, "Oldowan Tools," World History Encyclopedia, July 13, 2020, https://tinyurl.com/47hvzy4b.

11. Donald R. Prothero, *Evolution: What the Fossils Say and Why It Matters* (Columbia University Press, 2017), 366–67; Peter Andres and Christopher Stringer, "The Primates' Progress," in *The Book of Life*, ed. Stephen Jay Gould (Norton, 2001), 240–41.

12. "The Life and Times of Turkana Boy," Turkana Basin Institute, accessed August 27, 2025, https://tinyurl.com/4hz8hzbm;

"KNM-WT 15000," Smithsonian National Museum of Natural History, accessed August 27, 2025, https://tinyurl.com/45u69b7h.

13. Prothero, *Evolution*, 367–68; Steve Brusatte, *The Rise and Reign of Mammals: A New History from the Shadow of the Dinosaurs to Us* (Mariner Books, 2023), 385–86; Josie Glausiusz, "What Drove *Homo erectus* Out of Africa?" *Sapiens*, October 14, 2021, https://tinyurl.com/mry9h5np; Karie Pavid, "Human Ancestor *Homo erectus* Had the Stocky Chest of a Neanderthal," Natural History Museum, July 6, 2020, https://tinyurl.com/yj4rybjt; Lisa Hendry, "*Homo erectus*, Our Ancient Ancestor," Natural History Museum, accessed August 27, 2025, https://tinyurl.com/5xwtkdu8.

14. Holly Chetan-Welsh and Lisa Hendry, "*Homo floresiensis*: The Real-Life 'Hobbit'?" Natural History Museum, accessed August 27, 2025, https://tinyurl.com/y224rj3r; "*Homo floresiensis*," Smithsonian Museum of Natural History, last updated July 1, 2022, https://tinyurl.com/58nkvp7v.

15. Coyne, *Why Evolution Is True*, 207.

16. Laura T. Buck and Chris B. Stringer, "*Homo heidelbergensis*," *Current Biology* 24, no. 6 (2014): 214–15, https://tinyurl.com/bddscs89; Barbara Helm Welker, *The History of Our Tribe: Hominini* (Pressbooks, 2017), 206–15; Fran Dorey, "*Homo heidelbergensis*," Australian Museum, June 28, 2021, https://tinyurl.com/jfy3x68w; "*Homo heidelbergensis*," Smithsonian Museum of Natural History, last updated January 3, 2024, https://tinyurl.com/37cp8rkx.

17. Matthias Meyer and Sandra Jacob, "400,000-Year-Old Fossils from Spain Provide Earliest Genetic Evidence of Neandertals," Max-Planck-Gesellschaft, March 14, 2016, https://tinyurl.com/58wbwa23.

18. Simon E. Fisher, "Human Genetics: The Evolving Story of *FOXP2*," *Current Biology* 29 (2019): 65–67, https://doi.org/10.1016/j.cub.2018.11.047.

19. "*Homo neanderthalensis*," Smithsonian National Museum of Natural History, last updated January 3, 2024, https://tinyurl.com/2wd2rujv.

20. Christopher Joyce, "Study: Neanderthals Wore Jewelry and Makeup," NPR, January 12, 2010, https://tinyurl.com/4tyeskdd.

21. Frances Vinall, "Neanderthal Community Cared for Child with Down Syndrome, Fossil Suggests," *Washington Post*, June 27, 2024, https://tinyurl.com/5at7wy9r.

22. "*Homo heidelbergensis*," Smithsonian Museum of Natural History.

23. Brian Handwerk, "An Evolutionary Timeline of *Homo sapiens*," *Smithsonian Magazine*, February 2, 2021, https://tinyurl.com/2j4aju3b.

24. Guy Gugliotta, "The Great Human Migration," *Smithsonian Magazine*, July 2008, https://tinyurl.com/ytku32xe; Brusatte, *The Rise and Reign of Mammals*, 387–88.

25. Will Dunham, "Where Did *Homo sapiens* Go After Leaving Africa? New Study Has an Answer," Reuters, March 25, 2024, https://tinyurl.com/54zprhmd.

26. Gugliotta, "The Great Human Migration."

27. "Africans Have the Greatest Genetic Variation," CBS News, April 30, 2009, https://tinyurl.com/4wew9wv7.

28. "What Does It Mean to Have Neanderthal or Denisovan DNA?" Medline Plus, National Library of Medicine, last updated June 23, 2022, https://tinyurl.com/4kydfv2j.

29. Hugo Zeberg and Svante Pääbo, "A Genomic Region Associated with Protection Against Severe COVID-19 Is Inherited from Neanderthals," *Nature* 587 (2020): 610–12, https://doi.org/10.1038/s41586-020-2818-3; Laura Ungar and Maddie Burakoff, "More Research Is Examining How We Carry the 'Genetic Legacy' of Extinct Human Species," PBS, September 25, 2023, https://tinyurl.com/4f6pexpv; Emily Cooke, "10 Unexpected Ways Neanderthal DNA Affects Our Health," LiveScience, May 17, 2024, https://tinyurl.com/hf2au9jc.

30. "What Does It Mean to Have Neanderthal or Denisovan DNA?"

31. "The Mysterious Denisovans Have at Last Come in from the Cold," *Nature*, May 1, 2019, https://doi.org/10.1038/d41586-019-01310-7.

32. "Darwin and the Orangutan," London Zoo, accessed August 27, 2025, https://tinyurl.com/323pumzs.

CHAPTER 8

1. Ruth Coker Burks, *All the Young Men: A Memoir of Love, AIDS, and Chosen Family in the American South* (Grove, 2020).

2. Kate Kelland, "The Retroviruses," CEPI, May 15, 2024, https://tinyurl.com/yuhzdc3w.

3. "June 2000 White House Event," National Human Genome Research Institute, last updated August 29, 2012, https://tinyurl.com/62eyf3p6.

4. Laura Vargiu et al., "Classification and Characterization of Human Endogenous Retroviruses; Mosaic Forms Are Common," *Retrovirology* 13, no. 7 (2016), https://doi.org/10.1186/s12977-015-0232-y.

5. Adam Lee et al., "Identification of an Ancient Endogenous Retrovirus, Predating the Divergence of the Placental Mammals," *Philosophical Transactions of the Royal Society B* 368, no. 1626 (2013), https://doi.org/10.1098/rstb.2012.0503.

6. Cécile Voisset, Robin A. Weiss, and David J. Griffiths, "Human RNA 'Rumor' Viruses: The Search for Novel Human Retroviruses in Chronic Disease," *Microbiology and Molecular Biology Reviews* 72 (2008), https://doi.org/10.1128/mmbr.00033-07.

7. Ulrike Schön et al., "Human Endogenous Retroviral Long Terminal Repeat Sequences as Cell Type-Specific Promoters in Retroviral Vectors," *Journal of Virology* 83, no. 23 (2009), https://doi.org/10.1128/JVI.00858-09.

8. Nicole Grandi et al., "HERV-W Group Evolutionary History in Non-Human Primates: Characterization of ERV-W Orthologs in *Catarrhini* and Related ERV Groups in *Platyrrhini*," *BMC Ecology and Evolution* 18, no. 6 (2018), https://doi.org/10.1186/s12862-018-1125-1.

9. Here's a short video explaining ERVs in humans and chimps: Stated Clearly, "DNA Evidence That Humans and Chimps Share a Common Ancestor: Endogenous Retroviruses," YouTube, April 5, 2021, 12:06, https://tinyurl.com/yp67p4mj.

10. David Quammen, "How Viruses Shape Our World," *National Geographic*, January 14, 2021, https://tinyurl.com/249rd9dm.

11. Here are two short videos helpful in understanding human chromosome 2 fusion: Kenneth Miller, "Ken Miller Human Chromosome 2 Genome," YouTube, November 28, 2007, 3:28, https://tinyurl.com/2s498zuw; Gutsick Gibbon, "Human Chromosome 2 VS Creationism: Bite-Sized Busts," YouTube, November 23, 2022, 28:22, https://tinyurl.com/5zx4ukya.

12. Graeme Finlay, *Human Evolution: Genes, Genealogies, and Phylogenies* (Cambridge University Press, 2013), section 3.2.4.

13. John Hawks, "When Did Human Chromosome 2 Fuse?" John Hawks, August 31, 2023, https://tinyurl.com/yc28ytau.

14. Philip K. Allan, "Finding the Cure for Scurvy," U.S. Naval Institute, February 2021, https://tinyurl.com/3rc2s73x.

15. Guy Drouin, Jean-Rémi Godin, and Benoît Pagé, "The Genetics of Vitamin C Loss in Vertebrates," *Current Genomics* 12, no. 5 (2011): 371–78, https://doi.org/10.2174/138920211796429736.

16. Jerry A. Coyne, *Why Evolution Is True* (Viking, 2009), 71–72.

17. Yoav Gilad et al., "Human Specific Loss of Olfactory Receptor Genes," *PNAS* 100, no. 6 (2003): 3324–27, https://doi.org/10.1073/pnas.0535697100.

18. Coyne, *Why Evolution Is True*, 71.

19. Susumu Ohno, "So Much 'Junk' DNA in Our Genome," *Brookhaven Symposium in Biology* 23 (1972): 366–70.

20. Finlay, *Human Evolution*, section 3.3.

21. Mitch Leslie, "Yeast Can Live with Human Genes," *Science*, May 21, 2015, https://tinyurl.com/msbeas49.

22. Andrew Fabich, "Do Endogenous Retroviruses (ERVs) Support Common Ancestry?" Answers in Genesis, December 10, 2015, https://tinyurl.com/3ep6n6yt.

23. Todd Wood, "Microbes Continue: Retroviruses," *Todd's Blog*, October 19, 2009, https://tinyurl.com/3uxvfkfd.

24. Todd Wood, "The Truth About Evolution," *Todd's Blog*, September 30, 2009, https://tinyurl.com/kzbac6w9.

25. Jeffrey P. Tomkins, "Human Chromosome 2 Fusion Never Happened," Institute for Creation Research, April 30, 2020, https://tinyurl.com/4acfkt33; Anjeanette Roberts, "Questioning Evolutionary Presuppositions About Endogenous Retroviruses," Reasons to Believe, December 7, 2017, https://tinyurl.com/3mmyt389; "Vestigial Genes?" Answers in Genesis, August 5, 2006, https://tinyurl.com/3nubfc2m; Dominic Statham, "Heads I Win, Tails You Lose," Creation Ministries International, November 11, 2010, https://tinyurl.com/muwrnehj.

26. Gutsick Gibbon, "84% Chimpanzee: The Nonsense Creationism of Jeffrey Tomkins," YouTube, December 30, 2023, 2:16:24, https://tinyurl.com/4cc5xanf.

27. Jack Neel, "Candace Owens on Government Experiments, Eminem Diss, and Fake Dinosaurs," YouTube, July 21, 2024, 1:04:57, https://tinyurl.com/2s3a8zy9.

CHAPTER 9

1. Anthony C. Allison, "The Discovery of Resistance to Malaria of Sickle-Cell Heterozygotes," *Biochemistry and Molecular Biology Education* 30, no. 5 (2002): 279–87, https://doi.org/10.1002/bmb.2002.494030050108.

2. "Human Genetics: Concepts and Application," PBS, accessed August 28, 2025, https://tinyurl.com/4r7tf5au.

3. "Malaria," World Health Organization, December 11, 2024, https://tinyurl.com/3dteep45.

4. Daniel Shriner and Charles N. Rotimi, "Whole-Genome-Sequence-Based Haplotypes Reveal Single Origin of the Sickle Allele During the Holocene Wet Phase," *American Journal of Human Genetics* 102 (2018): 547–56, https://doi.org/10.1016/j.ajhg.2018.02.003.

5. Tony Maccarella, "The Spread of Farming in Sub-Saharan Africa: Bantu Migration," OER Project, accessed August 28, 2025, https://tinyurl.com/2enkfwnw.

6. Nalina Eggert, "Tracing Sickle Cell Back to One Child, 7,300 Years Ago," BBC, March 12, 2018, https://tinyurl.com/mrxs6dj7; Carl Zimmer, "How One Child's Sickle Cell Mutation Helped Protect the World from Malaria," *New York Times*, March 8, 2018, https://tinyurl.com/54rmjp58.

7. Pascale Gerbault et al., "Evolution of Lactase Persistence: An Example of Human Niche Construction," *Philosophical Transactions of the Royal Society B* 366 (2011): 863–77, https://doi.org/10.1098/rstb.2010.0268.

8. Ewen Callaway, "How Humans' Ability to Digest Milk Evolved from Famine and Disease," *Nature* 608 (2022): 251–52.

9. "Got Lactase?" Understanding Evolution, April 2007, https://tinyurl.com/2spzap27.

10. "Himalayan Powerhouses: How Sherpas Have Evolved Superhuman Energy Efficiency," University of Cambridge, May 22, 2017, https://tinyurl.com/2tk6hdrz.

11. Hillary Mayell, "Three High-Altitude Peoples, Three Adaptations to Thin Air," *National Geographic*, February 25, 2004, https://tinyurl.com/yc3du2f7.

12. Ann Gibbons, "Tibetans Inherited High-Altitude Gene from Ancient Human," *Science*, July 2, 2014, https://tinyurl.com/yc68msxp.

13. Kelsey E. Witt and Emilia Huerta-Sánchez, "Convergent Evolution in Human and Domesticate Adaptation to High-Altitude Environments," *Philosophical Transactions of the Royal Society B* 374, no. 1777 (2019), https://doi.org/10.1098/rstb.2018.0235.

14. "Types of Colour Blindness," Colour Blindness Awareness, accessed August 28, 2025, https://tinyurl.com/m6suhtrc.

15. "Swyer Syndrome," National Organization for Rare Disorders, last updated September 18, 2019, https://tinyurl.com/mujh3v55.

16. Jenny Graves, "Sex, Genes, the Y Chromosome and the Future of Men," *Conversation*, November 14, 2014, https://tinyurl.com/wtrnaumu.

17. Jason Wilson, Joshua M. Staley, and Gerald J. Wyckoff, "Extinction of Chromosomes Due to Specialization Is a Universal Occurrence," *Scientific Reports* 10, no. 2170 (2020), https://doi.org/10.1038/s41598-020-58997-2.

18. Miho Terao et al., "Turnover of Mammal Sex Chromosomes in the *SRY*-Deficient Amami Spiny Rat Is Due to Male-Specific Upregulation of *Sox9*," *PNAS*, November 28, 2022, https://doi.org/10.1073/pnas.2211574119.

19. M. R. Hayden et al., "The Origin of Huntington's Chorea in the Afrikaner Population of South Africa," *South African Medical Journal* 58, no. 5 (1980): 197–200.

20. Elizabeth Pennisi, "Humans Are Still Evolving—and We Can Watch It Happen," *Science*, May 17, 2016, https://tinyurl.com/2nm6eddk.

21. "The Biology of Skin Color," HHMI BioInteractive, May 14, 2020, https://tinyurl.com/v9aferak.

22. "Your Aching Back: A Gift from Your Ancient Ancestors," PBS, April 9, 2014, https://tinyurl.com/44ubypv3.

23. "Anthropologist Explores Challenges for Humans Based on Evolution," Case Western Reserve University, February 19, 2013, https://thedaily.case.edu/anthropologist-explores-challenges-for-humans-based-on-evolution/.

24. Jonathan McLatchie, "A Leading ID Proponent Rebuts Our Anti-ID Challenges," September 22, 2023, in *Recovering Evangelicals*, podcast, host Luke Janssen, 1:08:46, https://tinyurl.com/59hk7rbu.

25. Bjarne Røsjø, "Evolutionary Flaws Disprove the Theory of Intelligent Design," University of Oslo, March 11, 2020, https://tinyurl.com/5yp44syf.

26. Michael J. Behe, *Darwin's Black Box: The Biochemical Challenge to Evolution* (Free Press, 2006), 223–24.

27. Steve Laufmann and Howard Glicksman, *Your Designed Body* (Discovery Institute Press, 2022), 441.

28. Psalm 139:14.

CHAPTER 10

1. Ewen Callaway, "Genetic Adam and Eve Did Not Live Too Far Apart in Time," *Nature*, August 6, 2013, https://doi.org/10.1038/nature.2013.13478; Dennis Venema, "Mitochondrial Eve and Y Chromosome Adam," BioLogos, March 13, 2014, https://tinyurl.com/a84mnyh2.

2. Stephen Schaffner, "What Genetics Says About Adam and Eve," BioLogos, July 11, 2021, https://tinyurl.com/4rxa2v2z.

3. Schaffner, "What Genetics Says About Adam and Eve."

4. Brian Thomas, "Does Modern Genetics Confirm a Historical Adam?" Institute for Creation Research, March 31, 2016, https://tinyurl.com/45jdruju.

5. Schaffner, "What Genetics Says About Adam and Eve."

6. Schaffner, "What Genetics Says About Adam and Eve."

7. Jeff Tomkins, "Out of Babel—Not Africa: Genetic Evidence for a Biblical Model of Human Origins," *Journal of Creation* 34, no. 1 (2020): 79–85, https://tinyurl.com/24jb8vxe; "The 10 Best Evidences from Science That Confirm a Young Earth," Answers in Genesis, November 13, 2021, https://tinyurl.com/4fj2262u; "Biblical Age of the Earth," Institute for Creation Research, accessed March 8, 2025, https://tinyurl.com/yhnnshtt.

8. Thomas, "Does Modern Genetics Confirm a Historical Adam?"; Nathaniel T. Jeanson, "Getting Enough Genetic Diversity," Answers in Genesis, July 23, 2016, https://tinyurl.com/32d276fh; Robert W. Carter, "Adam, Eve, and Noah vs.

Modern Genetics," Creation Ministries International, March 7, 2010, updated May 11, 2010, https://tinyurl.com/ys4kh2fn.

9. John C. Sanford et al., "Adam and Eve, Designed Diversity, and Allele Frequencies," *Proceedings of the International Conference on Creationism* 8 (2018): 200–216.

10. Carter, "Adam, Eve, and Noah vs. Modern Genetics"; Jonathan Sarfati, "Eve, the Rib, and Modern Genetics," *Creation* 40, no. 2 (2018): 44–47.

11. Fazale Rana, *Who Was Adam? A Creation Model Approach to the Origin of Humanity* (Reasons to Believe, 2015), 79.

12. Leslie A. Pray, "Genetic Drift: Bottleneck Effect and the Case of the Bearded Vulture," *Nature Education* 1, no. 1 (2008): 61, https://tinyurl.com/5xntryyt.

13. Alan R. Templeton et al., "Disrupting Evolutionary Processes: The Effect of Habitat Fragmentation on Collared Lizards in the Missouri Ozarks," *PNAS* 98, no. 10 (2001): 5426–32, https://doi.org/10.1073/pnas.091093098.

14. Population bottlenecks occur when a population's size is reduced dramatically for at least one generation. Common causes of bottlenecks are disease, environmental disasters, and overhunting.

15. Rana, *Who Was Adam?*, 79.

16. Ola Hössjer and Ann Gauger, "A Single-Couple Human Origin Is Possible," *BIO-Complexity* 1 (2019): 1–20, http://dx.doi.org/10.5048/BIO-C.2019.1; Ann Gauger, "New *BIO-Complexity* Paper: We Could Have Come from Two," Science & Culture Today, October 21, 2019, https://tinyurl.com/yk9y5zab.

17. Hans Ellegren and Nicolas Galtier, "Determinants of Genetic Diversity," *Nature Reviews Genetics* 17 (2016): 422–33, https://doi.org/10.1038/nrg.2016.58; Jeffrey D. Wall, "Estimating Ancestral Population Sizes and Divergence Times," *Genetics* 163 (2003): 395–404.

18. Fazale Rana, "Conservation Biology Studies Elicit Doubts About the First Human Population Size," Reasons to Believe, April 26, 2017, https://tinyurl.com/5n86jywp.

19. Georgia Purdom, "Were Adam and Eve Real People?" in *How Do We Know the Bible Is True?*, ed. Ken Ham and Bodie Hodge (Master Books, 2012), 2:229–40.

20. Elizabeth Mitchell, "Did We All Come from Adam and Eve?" Answers in Genesis, December 19, 2013, https://tinyurl.com/3btmsdhc.

21. Denis Alexander, "The Various Meanings of Concordism," BioLogos, March 23, 2017, https://tinyurl.com/mvyzmjv4.

22. S. Joshua Swamidass, *The Genealogical Adam and Eve: The Surprising Science of Universal Ancestry* (IVP Academic, 2021).

23. Swamidass, *The Genealogical Adam and Eve*, 9–13.

24. William Lane Craig, *In Quest of the Historical Adam: A Biblical and Scientific Exploration* (Eerdmans, 2021). Hereafter, page references from this work will be given in parentheses in the text.

25. Melissa Cain Travis, "William Lane Craig Explores the Headwaters of the Human Race," *Christianity Today*, October 2021, https://tinyurl.com/mpmjjkap.

26. Dennis Venema and Scot McKnight, *Adam and the Genome: Reading Scripture After Genetic Science* (Brazos, 2017).

27. Venema and McKnight, *Adam and the Genome*, 51–55; Dennis R. Venema, "Genesis and the Genome: Genomic Evidence for Human-Ape Common Ancestry and Ancestral Hominid Population Sizes," *Perspectives on Science and Christian Faith* 62, no. 3 (2010): 166–78.

28. Dennis Venema, "Adam—Once More, with Feeling," *Patheos*, Jesus Creed, last updated November 4, 2019, https://tinyurl.com/mr44auap.

29. Venema and McKnight, *Adam and the Genome*, 55.

CHAPTER 11

1. Peter Enns, *The Evolution of Adam: What the Bible Does and Doesn't Say About Human Origins* (Brazos, 2012), 38.

2. Enns, *The Evolution of Adam*, 47.

3. Enns, *The Evolution of Adam*, 20–26; see also "Documentary Hypothesis," University of Pennsylvania School of Arts & Sciences, accessed September 2, 2025, https://tinyurl.com/2nckn69t.

4. Walter Brueggemann, *Theology of the Old Testament: Testimony, Dispute, Advocacy* (Fortress, 1997), 74–75.

5. "A Modest Proposal" is an eighteenth-century essay by Jonathan Swift. Swift exposes the excesses of the elite and the plight of the poor by suggesting the poor sell their children as food for the rich.

6. J. Richard Middleton, *The Liberating Image: The Imago Dei in Genesis 1* (Baker Academic, 2005), 264–65.

7. Enns, *The Evolution of Adam*, 71–72.

8. Isaiah 66:1–2.

9. Enns, *The Evolution of Adam*, 72.

10. John H. Walton, *The Lost World of Genesis One: Ancient Cosmology and the Origins Debate* (IVP Academic, 2009); John H. Walton, *The Lost World of Adam and Eve: Genesis 2–3 and the Human Origins Debate* (IVP Academic, 2015).

11. Walton, *The Lost World of Genesis One*, 91.

12. Walton, *The Lost World of Genesis One*, 87–92.

13. Walton, *The Lost World of Genesis One*, 72–77.

14. Middleton, *The Liberating Image*, 127–28.

15. Genesis 2:7.

16. Middleton, *The Liberating Image*, 25.

17. Middleton, *The Liberating Image*, 94.

18. Daniel 3:1–30.

19. Walton, *The Lost World of Adam and Eve*, 41–42.

20. Psalm 8:5–8.

21. Walton, *The Lost World of Adam and Eve*, 105–6.

22. Walton, *The Lost World of Adam and Eve*, 105–8.

23. Middleton, *The Liberating Image*, 284.

24. Ryan S. Peterson, "The Imago Dei as Human Identity: A Theological Interpretation," *Journal of Theological Interpretation* Supplement 14 (Eisenbrauns, 2016), 1–2.

25. Middleton, *The Liberating Image*; Walton, *The Lost World of Adam and Eve*.

26. Richard Middleton, "Image of God," May 7, 2020, in *Language of God*, podcast, host Jim Stump, 52:09, https://tinyurl.com/bdf7ucb2; Walton, *The Lost World of Adam and Eve*, 42.

27. Walton, *The Lost World of Adam and Eve*, 33.

28. Walton, *The Lost World of Adam and Eve*, 43.

29. Ken Ham (@aigkenham), "Creation week was a period of six ordinary 24-hour days. (Yes, we start with Genesis 1–11 when

it comes to our worldview. It's the foundation!)." X, November 30, 2024, 4:21 a.m., https://tinyurl.com/ysn7sarp.

30. Jason Yates, "A Conversation with Ken Ham: Why Understanding Genesis Is Foundational to the Christian Worldview," My Faith Votes, August 16, 2022, https://tinyurl.com/2sfyv4tf.

31. Scot McKnight, "How Genetic Science Affects the Bible," January 12, 2023, in *Holy Post*, podcast, hosts Phil Vischer and Skye Jethani, 52:40, https://tinyurl.com/2fd2u55u.

32. Romans 5:12.

33. McKnight, "How Genetic Science Affects the Bible."

34. Joseph A. Fitzmyer, *Romans*, Anchor Bible 33 (Doubleday, 1993), 410.

35. *In Memoriam A. H. H.*, published by Alfred, Lord Tennyson in 1850.

CHAPTER 12

1. Here's an easy-to-read and informative account of Galileo and the resistance to evidence: Bobby Valentine, "Some Looked and Couldn't See: Galileo, Seeing and the Quest for Truth," *Stoned-Campbell Disciple*, December 4, 2010, https://tinyurl.com/55mrbcbd.

2. Living Waters, "Rescued Astronaut Butch Wilmore Preaches the Gospel While Stuck on the ISS," Facebook, March 19, 2025, https://tinyurl.com/49325yeh.

3. Jim Stump, *The Sacred Chain: How Understanding Evolution Leads to Deeper Faith* (HarperOne, 2024), 80.

4. Stump, *The Sacred Chain*, 84.

5. Prosanta Chakrabarty, *Explaining Life Through Evolution* (MIT Press, 2023).

6. Prosanta Chakrabarty, "Ask an Expert: An Evolution Education," August 4, 2023, *Science Friday*, podcast, host Ira Flatow, 33:13, https://tinyurl.com/2zc8n6h8.

7. "Doubt & Faith: Top Reasons People Question Christianity," Barna, March 1, 2023, https://tinyurl.com/c6esw4n8.

8. Ian McConnell, "Important," YouTube, September 8, 2022, 02:10, https://tinyurl.com/mr9azc65.

9. Edward J. Larson, "The Biology Wars: The Religion, Science and Education Controversy" (Pew Forum's biannual Faith Angle

Conference on religion, politics, and public life, Key West, FL, December 5, 2005), http://pewrsr.ch/14gq8Mz.

10. Luke Janssen, "My Liberal Christian Worldview," March 7, 2025, *Recovering Evangelicals*, podcast, host Luke Janssen, 57:17, https://tinyurl.com/4uz26vpb.

INDEX